P9-EML-506

Insatiable Curiosity

Inside Technology

edited by Wiebe E. Bijker, W. Bernard Carlson, and Trevor Pinch

A list of books in the series appears at the back of the book.

Insatiable Curiosity

Innovation in a Fragile Future

Helga Nowotny

Translated by Mitch Cohen

The MIT Press
Cambridge, Massachusetts
London, England

For information about special quantity discounts, please email <special_sales@mitpress.mit.edu>.

This book was set in Stone Serif and Stone Sans by SNP Best-set Typesetter Ltd., Hong Kong. Printed and bound in the United States of America.

Printed with the support of the Austrian Ministry of Science and Research in Vienna.
Printed with the support of Austria-Cooperation, Vienna.

Library of Congress Cataloging-in-Publication Data

Nowotny, Helga.
[Unersättliche Neugier. English]
Insatiable curiosity : innovation in a fragile future / by Helga Nowotny; translated by Mitch Cohen.
p. cm.—(Inside technology)
Includes bibliographical references.
ISBN 978-0-262-14103-1 (hardcover : alk. paper)
1. Science—Social aspects. 2. Technology—Social aspects. 3. Science—Technological innovations. 4. Technological innovations. 5. Curiosity. 6. Creative ability in science. I. Title.
Q175.5.N68513 2008 303.48′3—dc22 2007048462

10 9 8 7 6 5 4 3 2 1

For Isabel and Gideon

With thanks to the Riksbanken Jubileumsfond in Stockholm, Sweden, for funding the translation from the German.

Printed with gratefully acknowledged support from the Austrian Federal Ministry for Science and Research and the Verein Oesterreich-Kooperation (Austria Cooperation Association), both in Vienna, Austria.

Contents

Insatiable Curiosity

1 The Emergence of the New

We cannot rely on nature to impose its own limits.
—Marilyn Strathern

Curiosity and the Preference for the New

A young friend who works with the deaf described a telling difference in their languages. In the sign language of the Euro-American world, the sign that stands for the future points to the front, and probably everyone in Europe and America, deaf or not, would give this direction as the one in which we all think the future lies. Not so in Africa, where the gesture points backward. What lies in front of us, according to the African explanation, is the past because only it is already known. The future, by contrast, lies where we cannot see it—behind us or around us.

The future cannot be equated with the new, which can also be discovered in the past. The new differs from the old and yet must resemble it enough to make the difference recognizable. The difference thus created allows what is new to link with what already exists. New and old can exist beside each other, replace each other, or enter into connections in which what exists appears unfamiliarly new or what is new seems known. The new must be brought into the familiar world and enter into exchange with prior experiences. It must be given meaning and evaluated. The new must be different, but to be recognizable as the new, it requires observers to make a concentrated effort.

However we want to locate it spatially, the future lies temporally in front of us, embedded in the biological processes that follow the arrow of time from birth to death. All societies distinguish between segments of time categorized as past, present, and future. These temporal structurings are subject to historical changes and are components of cultural cosmologies. The temporal horizon separating the future from the present can appear as a hard-edged, abrupt line at the boundary between chaos and order or as a narrow gap through which it is possible to enter an eternity removed from change. The future can be conceived as a smooth transition or as an "extended present" with the open future horizon that entered history with the European modern period. With it arose for the first time the feeling of acceleration that is connected with the extent of the changes and the increased appearance of the new. The future itself, however, cannot be reached any faster. It cannot be overtaken and is in fact, just as the Africans regard it, around us, in front of us, and behind us all at the same time. The future *is*. Its content, its shape, and its fullness—the images we construct of it—always have significance only in the here and now.

But where does the fascination of the new come from? Everyone wants to know the future, especially her own, to be safe from unwanted surprises and to be able to at least partly master the unknown, which is always also a potential threat. But the desire to control the future seeks to protect what one already has and what one has achieved. The fascination with the new, by contrast, is activated by curiosity and the desire to explore the unknown. This curiosity induces us to take the next step that leads beyond familiar terrain. However tentative, cautious, or inexperienced this step may be, it goes wherever longing and the discovery of one's own latent wishes and desires may lead. The thin line separating the present from the future is irrevocably crossed. Curiosity aims to explore a space that must still be furnished for us. With questions and gestures more spontaneous than goal-oriented, curiosity explores what it does not yet know and what seems interesting and worth knowing, often for reasons it cannot name. It actively strives to hone itself on reality and to gain experience that gives reality a clearly perceptible form that can be interacted with. To gain this experience, curiosity uses all the senses and means available to human beings. It is insatiable in two ways: first, because the space of possibilities and reality that is to be explored still approaches infinity; and second, because more and more means and instruments, mostly but not entirely scientific and technical in nature, are at our disposal to expand the space of our experience.

The experiences triggered by curiosity, which are often based on trial and error, are an important reservoir and cultural resource for individually and collectively imagining the future. In connection with the economic development of the third world, the anthropologist Arjun Appadurai sees the conventional definition of culture as a decisive hindrance limiting the scope people urgently need to shape their lives. Culture is usually set in relation to the past and serves to preserve our legacy and tradition. But economic growth and development are associated with the future—with plans, hopes, and goals. In the anthropological understanding, the future usually has no place in models of culture. The science of the future is economics. It assumes that people have preferences and desires, and it models their expectations and their calculating behavior.[1]

But the ability to imagine the future—indeed, the desire for an imagined future—is a cultural ability inherent in all people, including those who, because of their miserable economic circumstances, supposedly have no future. The capacity to aspire, as Appadurai calls it, cannot be reduced to individual preferences and the field of markets but grows out of cultural norms and values. While the affluent have a greater range of experiences at their disposal, know their own wishes and aspirations better, and know the means that can be used to achieve the latter, they are usually also in a more advantageous position to try out new experiences and to implement them purposefully. The capacity to aspire puts them in a better position to navigate an unknown future. This ability to claim the future for oneself is a cultural resource and should be made available also to those who currently do not have it, like the poor in the developing countries.

But it is also a cultural resource potentially available to all who currently feel overwhelmed by the plethora of innovations and the speed with which they are created and introduced. Fear of the future arises from the feeling of losing control over how one leads one's life. It suppresses curiosity and narrows the scope of experience. It reduces one's possibilities for trying out the new. Neither the deep insecurity that accompanies this nor the experience of being steamrollered by events and developments is historically new. The nineteenth century was convulsed by the effects of the industrial revolution and lived through a previously unknown wave of revolutions that "dissolved everything solid into air," as Marx put it. Psychiatrists diagnosed new syndromes like neurasthenia, which they traced to the failure of the human organism and especially of the nervous system to keep up with these changes.

Subtle literary testimony and astute observations still give us today a taste of the intensity of the emotions at that time—for example, when electric lights, the telegraph, and the railroad enter the world of Marcel Proust's literary characters, when Marcel makes his first telephone call, and when he experiences the arrival of his first automobile. He sees these as the great "compressions" of social life, as various communities begin interlocking and become more compact and as the isolation of small villages like Combray is overcome. Everyone is subjected more intensely to social pressure than they were before, which leads to greater conformity and greater anxiety about the authenticity of the self. It is as if this accelerating physical and communicative contact with others does not so much foster as suspend communicative closeness and the intimacy of daily life—the feeling that others see one, the feeling of constant surveillance. Since there are indeed fewer places where one is not seen or can hide oneself, there are also fewer reasons to avoid contact. There are fewer occasions to be alone and to develop a self that is independent of society.[2]

The dependencies that today obscure the view of an imaginary future bear the stamp of globalization, that inscrutable network of world domination and markets, the outsourcing of jobs, and the increase in worldwide competitive pressure. Behind these are structural shifts and fractures that accompany the advance of neoliberalism, on the one hand, but also the shift from classical industrial production to the conceptual and knowledge industries, on the other. Scientific and technological knowledge and the products and infrastructures they bring forth thus have central importance. They are regarded as the driving force for continued economic growth and as indispensable in achieving decisive competitive advantages.

The map available for navigating this sea of opportunities and dangers is confusing because, depending on position and experience, it offers different starting positions for realistically exploring the future. For some, the space of the future is filled with new technological visions and highly promising mini-utopias that hold the potential to make life easier, better, and more beautiful. For others, the horizon of the future is darkened with dystopias. At stake is the maintenance of one's own identity, whether endangered by a cultural diversity that is seen as a threat or by a step-by-step loss, felt as a confiscation, of control over one's own life through growing dependence on technology and scientifically generated innovations. The circumstances under which something new can be tried out, driven by and in playful association with curiosity, then dwindle. The possibility of having new experiences and of encountering a changing and emerging reality with practices that permit trial and error and the exploration of one's own wishes and their implementation also dwindle. The capacity to aspire suffers under this.

These opposing processes make it more difficult for people today to conceive the future and to develop clear ideas about their own scope of influence. This explains the fateful oscillation between regression to the hubris of blind faith in progress that so calamitously characterized the twentieth century and the temptations of a fundamentalism that clutches at fragile and false securities to avoid encountering the future. Conceiving the future—conceiving it differently—demands that we escape the polarities of utopias and dystopias and replace them with other images that are neither taken directly from science fiction nor fueled by media-staged apocalyptic or superhuman fantasies. Conceiving the future demands knowledge and imagination, a shifting back and forth between seriousness and play, science and irony. Knowledge must be spanned widely and, like the convergent technologies much praised today, must strive for an integration that draws from all available sources—the humanities as well as the natural and engineering sciences, the arts as well as technology and the experience of simple everyday life. Conceiving the future means examining the assumptions on which it supposedly rests. The inextricable and confusing bundle of forces and processes—of institutions and power relations in which the insatiability of curiosity encounters the diverse possibilities of its realization and implementation in the framework of a globalized capitalism and the political disorder of the world—must be seen as what we make of it culturally: a continuation of modernity.

Of course, modernity takes on new accents and fractures. In a situation characterized by loss and failure as well as by the striving to fill the emptiness of the future, to constrain it to a forming will, and to try out the freedoms that its possibilities promise, the historically socialized glance and the language of gestures are still forward-oriented. And yet new uncertainties are mixed in with the old, familiar language of gesture and its meanings. Uncertainties appear that result from the innovations with which science and technology open up a world that is different from everything previously thought, known, and seen. We are as able to see into the inside of our bodies using imaging techniques as we are to look through telescopes back millions of years to the primal history of the universe. Modern medicine extends the human life span, and yet worldwide epidemics still repeatedly break out and are still hard to stem. While some already dream of immortality, millions of people die for lack of basic medical care.

But the new that promises the future has a name. It suggests too much and too little at the same time and is as elusive and vague as it is demanding and determined. It is based on a fundamental societal consensus that is nevertheless brittle and must be constantly renegotiated. The name of the new is *innovation*. The word is often used in a deceptively simple form to mean a preference for the new. But what is worth striving for is not the new in general or innovations for their own sake or even the "mysterious banality" of fashion. As a typical phenomenon of modernity, innovation is contingent but not arbitrary. It replaces the unambiguous, traditional order with an unstable equilibrium in which the stability is the result of a demanding connection between various instabilities and is no longer their prerequisite. Fashion is an interplay between social contingency (everyone wants to be original and just like the rest) and temporal contingency (every present appears new and different due to the prerequisite of a past that permits it to be perceived in this way). Fashion plays with chance, which can be neither mastered nor foreseen, and is nonetheless able to operationalize it. It creates reference points and models that are to be deviated from to realize one's own original variants. The model is used to construct an identity of one's own by means of deidentification.[3]

Innovation creates instead the impression that it is the new, state-of-the-art navigation map that offers orientation on the uncertain journey into a fragile future. Driven by the capacity to aspire, it does not predetermine either content or goal. Instead, it promises to provide new experiences that must measure themselves against and hone themselves on an equally changing reality to lead to robust results. Innovation reminds us that the possibility of failure is always on board; it nonetheless encourages us to continue the journey. It plays with coincidence and the attempt to instrumentalize coincidence. It strives to increase the diversity of new forms because that is the only way the new can arise outside an already determined space of possibilities, the only way that, without wanting to predetermine the new, it can extend its effect beyond the process of arising by leading to further innovations. The new combination of already known or existing components, which Joseph Schumpeter said determines the process of innovation, points in the direction of a diversity with the potential to become ever greater. For the more innovations there are, the greater is the number of components from which new combinations can be produced in a rapidly growing process of combinatorics, without the contents being foreseeable and without categories for their description already existing. But the contexts of application must also multiply to offer the diversity of new forms the space of possibilities in which innovations not only arise but can also stabilize, solidify, and materialize. The "essential new" can be neither anticipated nor described, but it requires enough empty places where it can dock and a future that is empty enough to be open for the capacity to aspire but cannot be pinned down to its fulfillment. And yet—such is the law of the new—this future must be different than the present. There must be a clearly recognizable difference from what already exists.

Difference and Diversity

From the standpoint of evolutionary theory, sexuality can be understood as a machinery that creates differences. This is a biologically tested and proven method of creating the new. The advantages are numerous. A population that reproduces sexually can develop more rapidly than is the case with asexual organisms. Its descendants have a greater diversity of phenotypes, and in the short term, this diversity promises greater chances of adapting to changed environmental conditions. In the long term, it is an insurance against the unforeseeable since it increases the number of options and thereby the chances of survival.

Human societies have invented cultural machineries for creating differences. The experience and assertion that something is new, says sociologist Niklas Luhmann from the perspective of systems theory, mark the decision to use previously redundant possibilities to create structures. They are nothing other than an aspect of the system's self-description. That's why this change in the self-description underscores discontinuity to deconstruct traditions and to be able to reorganize connectivity.[4] The change appears on the meta level in the "difference that makes a difference." The new thus always implies a relation to the existing system; it maintains a relationship to the old.

The cultural machinery that creates differences functions by consciously or unconsciously drawing boundaries between the old and the new. Depending on how these distinctions are set, the new can appear as something whose contours are already known or as a radical break with the given. The less foreseeable the new is, the more it overtaxes perceptual and descriptive competencies. The new appears in two variants. First, it presents itself as a recombination of already existing and thus known elements—as a more or less continuous further development of the existing that pushes forward on the temporal axis into the future. The second variant is discontinuity—the break that brusquely underscores the contrasts to the existing and the ways that the new is different in thinking, seeing, doing, and living.

In the myths of origin—those ideas of the world and the forces reigning in it that almost all early societies developed—these two initial strands appear in the emergence of the new. Either the world and humanity begin with an act of creation (whoever the creator may be and whatever quality this act may have), or the myth of origin takes recourse to iterative processes that generate and regenerate themselves without a clear beginning. Today we are in the process of generating a third myth of origin—that of the scientific-technological civilization that constantly produces innovations out of itself. These come from unexpected scientific-technological breakthroughs or emerge as answers to societal demands for new solutions to problems. This scientific-technological myth of origin posits that the new beginning is constantly repeated and yet is different each time. The origin is the process of innovation itself, a process that has prerequisites but that, thanks to scientific-technological curiosity, continues to create out of itself.

Despite the seemingly arbitrary setting of boundaries and distinction, the definition of the new is never random. The perception and the need to describe the new demand cutoffs and distinctions so that quantum jumps and marked transitions can be recognized. In this way, the new can be distinguished from what already exists and from what arises elsewhere. The different strands of conceiving the beginning are carried forward in the processes of continuity and discontinuity, which together constitute the interweaving of the texture of life. Although there is no compelling reason why evolution must increase complexity, this is what we observe in biology.[5] The increase in complexity depends on a small number of large transitions in how genetic information is passed on between generations. Some of these transitions are unique, like the transition from the prokaryotes to the eukaryotes (now also called archaea and bacteria) and the emergence of the genetic code. Others, like the origin of multicellularity and animal societies, arise several times and independently of each other. There is no doubt that the evolutionary transition from ape to man correlates with an increase in cognitive faculties, which were again increased with the acquisition of speech competence. We already find among the primates a connection between the size of the brain and the complexity of the social system. The origin of language is still one of the most fascinating topics of research.

Similarly, the cultural machinery of innovation produced a number of marked transitions. They led to an increase in complexity, combined with an increase in social abilities to cope with the consequences—that is, to process them and turn them to productive use. Again and again, the flash of creativity manifesting itself individually or in small groups in art or science formed the basis for starting and crystallization points for innovations. In retrospect, the societal or economic conditions that led to heightened creativity can be reconstructed, but only very general statements about the emergence of creativity can be made.

With the beginning of the modern age, the production of the new was delegated primarily to one institution. Modern science appeared beside technology, which had long been independent of it. Since its institutionalization in the seventeenth century, science has specialized in the production of new knowledge and in discoveries. Combining different pieces of existing knowledge produces new knowledge the same way that putting together existing technological components leads to new technological inventions and ultimately to technological systems. Phenomena have long since been produced that do not appear in nature (most recently, synthetic genes). What nature organically shows us can be done is imitated with increasing success and inexorably helps open up invisible areas on the molecular level. What can be converted into information becomes information and can be accordingly processed. The convergent technologies based on successful connections among the biological, informational, nano-, and cognitive sciences open up a broad field in which brain and matter, body and environment can interact in a controlled fashion. These and other transformations that spring from science and technology touch on humanity's self-understanding as much as they change our social life together. The ensuing public debates oscillate between technical utopias and social dystopias. The first celebrate new possibilities of application and of applied knowledge and promise an arsenal of technological fixes for problems currently insoluble. The second point to the destabilizing potential that threatens human life together and that lament the loss of freedom and the delegation of responsibility to an electronically controlled world.

In the short period of the last three or four hundred years, a given, divinely created world transformed into a possible and enabling world whose discovery, invention, and recombination are owed to new ways of looking at things, technologies, and natural-scientific and technical mechanisms of explanation, access, and control. The myths still conveyed a certain idea of the world and its powers that promised security. They provided an image of the only possible world in which every sign and every unforeseen event could be interpreted to correspond with the overarching order and confirmed the corresponding view of the world. With the scientific interpretation of the world, the space of possibilities began expanding, and the possible world began multiplying. Here is where we find the development of the ambivalence so characteristic of modernity, which is expressed not only toward science and technology.

A concept of security and of a rationality promising security that is limited to a Cartesian viewpoint cannot keep pace with the multiplication and expansion of possibilities. Stephen Toulmin distinguishes between two forms of rationality. One is the Cartesian, which claims for itself a monopoly on knowledge and asserts that it knows the sole access through reason. The counterposition is represented by the enlightened skeptic Montaigne, who couples his belief in reason with skepticism, which is to be applied to both the questions and the answers. These two sides of rationality are two forms of reflexivity that stand in a field of tension in all modern societies. The point is thus not the confrontation between the Enlightenment and fundamentalism—between the religious and the secular understanding of the world—but a contradiction inherent in the Enlightenment itself. An exclusive monopoly on interpretation is claimed in the name of the natural sciences, against which stand other interpretative claims in the names of other complementary but also contradictory forms of knowledge.[6] If the first, cool variant of reason still bets on a mostly deterministic reality solely reacting to "facts," then the skeptical variant opens a space for the imagination and for a subjectively experienced reality that emotions and aesthetics help shape.

The "scientific method" that marks the natural-scientific monopoly on interpretation is, in reality, a bundle of extremely disparate methods, experimental and nonexperimental approaches that change and develop further over history. It provides precise, usually mathematically formalized possibilities of depiction and of controllable intervention and makes it possible to liberate the newly discovered or newly generated from the suspicion of being false or deviant. To speak with Luhmann, this is how the "true/false" code entered science. In a long-lasting process of emergence and refining, the struggle was over the "facts," which alone stood for reality, as described and explained by the natural sciences. The aim was to "cleanse" the facts to be able to put them on a solid foundation of proof that is stripped of their original context and thus generally valid. Only then could they be separated from "values"—from the wishes, feelings, and capacity to aspire that repeatedly threatened to contaminate the facts again.

Since the new is, on principle, open and includes everything that stimulates curiosity (which was increasingly conditioned and socialized and whose methods were refined), the scientific procedure seeks to continue posing questions. It does not primarily serve to solve problems but to develop them further. Grasping a problem—posing a question in a manner that makes it possible to tap new dimensions that take the problem further—is often considered the fruitful beginning of a new research program. The horizon of knowledge is as open as the horizon of the future. In this phase, the capacity to scientifically aspire takes its full effect. This is science in the making, the still open, preliminary process of research.

This principled attention and openness toward a realm of the possible, destined to be reduced to become actualized, aims to change the prior possible into the later factual. This also explains why scientific knowledge has the status of being provisional knowledge. Rational procedures that serve to test the various possibilities and that themselves are considered scientifically secured create certainties, but they remain reliable knowledge only for an indefinite period. The cultural dynamics of modern science are based on this movement forward. Opening a wide variety of possibilities begins their—temporary—selection for probable or actual givens. Even these operations are never brought to a final conclusion. The dissolution of the visible world into its invisible components, which are made visible by the use of instruments and image-giving technologies, opens up another growing space of knowledge. Or as Luhmann says, "Dissolution and recombination are conducted as a unity, and this unity, as much as the comparison, is a condition for the appearance of *new* knowledge, i.e., for acquiring knowledge. It is thereby necessary to want what is at the same time unwanted: the increasing probability of uncontrollable recombinations."[7]

The Taming of Curiosity

Wanting what is not already wanted, controlling what is still unforeseeable: these are the problems that move research and the public today, though in different ways. If not everything that is scientifically possible can or should be realized, what criteria of selection should be applied, and what societal orientation is there for the production of what does not yet exist? The social order strives for at least a minimum of societal continuity and foreseeability. The striving for consensus increases the pressure on the selection in another way—not having to accept everything that appears to be technically and scientifically feasible or not accepting all technological visions. At the top of the list of the fears that are publicly articulated today is the threat of loss of control over oneself and over how one leads one's life.

Today, the advance of liberal democracies and neoliberal economic systems has led to the celebration of what is regarded as its foundation, the autonomy of the individual, which is the result of centuries of struggle for liberation from political domination and religious censorship, at the same time as the neurosciences have cast doubt on whether the assumption of free will can be justified. And while consumers are subject to the suggestion that their decisions in purchasing products is freer, more informed, and more independent than ever before, in one area of research called *neuroeconomics*, imaging techniques like PET and fMRI are coupled with sophisticated experiments in behavioral economy to find out how purchasing decisions are actually made and how they can be influenced.

Today, the cultural-historically unique preference for the acquisition of new scientific knowledge is coming to a socially explosive head. On the one hand, the aim is to keep the machinery that creates differences and brings forth the new running efficiently—indeed, to enhance it. On the other hand, it is turning out that the innovation machinery has a number of societal blind spots. Aside from existing interests, the expanded space of possibility itself must first be explored. Most of the effects on society are not known since technological-scientific innovations also presuppose social innovations and depend on them for their success. For this reason, a balance is needed that guarantees the necessary degree of societal orientation, on the one hand, and that can produce a sufficiently dynamic preliminarity, on the other. But how can the accompanying instability in relation to the "play of possibilities" be accepted by society? How can a basic societal consensus be found that affirms and accepts the unforeseeability that is inseparable from research?

For as François Jacob aptly put it, "What we can suspect today will not become reality. There will be changes in any case, but the future will be different from what we think. That is especially true for science. Research is an endless process about which one can never say how it will develop. Unforeseeability is part of the essence of the venture of science. If one encounters something really new, then by definition this is something that one could not have known in advance. It is impossible to say where a particular area of research will lead." And he adds, "One must also accept the unexpected and the disquieting."[8]

Between society's preference for the new (as expressed in the institutionalization of modern technosciences and their societal status, which saw several centuries of a process of cultural forerunners, diverse revaluations, and many-layered revolutions in the structure of society) and a publicly articulating civil society that now presses for additional selection criteria in the process of acquiring knowledge, a zone of uncertainty is emerging and growing. It is essentially based on the fact that all knowledge that produces the new expands without itself being able to provide the criteria in accordance with which it can be limited again. The greater the desire for the unexpected that is brought forth by research in the lab, the more the pressure of expectation grows to bring it under control and steer it in specific directions while excluding other directions. The aim is to tame scientific curiosity and yet to give it free rein. The pressure for this comes from two seemingly opposite directions.

The taming of scientific curiosity takes place in the public space and is initiated by changes that are visible in a broader framework. One form of this taming takes the direction of a privatization of knowledge, or more precisely of the increasing tendency to register and exercise rights of ownership and disposition over scientific knowledge, data, methods, and new forms of life or organisms created in the laboratory. Perhaps researchers themselves unconsciously contribute to this when, seeking to protect their legitimate interests, they register ownership rights to their findings and see themselves and behave more like knowledge owners than like knowledge workers.[9]

Behind this development is the shift in the increasing level of investment in research from the state to the private sector. With this, a regime moves into research that has already successfully dominated the industrial and service sectors and that is now also supposed to efficiently cover the rising need for investment in science and guarantee greater efficiency in producing knowledge. Intellectual property rights, patents, licenses, and similar arrangements aim to ensure that competitive thinking and a greater proximity to the market lead to a more rapid transfer from the laboratory to marketable products or new technological systems. The trend to the privatization of knowledge is thus part of a broader pattern of societal changes. If the growing need for funding is to be covered by the private sector while the public research budget stagnates, then intellectual property rights will expand, and this will inevitably change the way researchers work, including their relationships to each other. Equally, basic research, in which future applications are still mostly uncertain, will move closer to possible contexts of application.

The second tendency is for civil society to have a greater voice in decision-making processes that center on complex scientific-technological matters and that are expected to have far-reaching effects on people's lives. If, in the first tendency, the shift in financing was the starting point for increasing privatization and "propertization" of scientific curiosity, then the second tendency has to do with the process of democratization, which does not stop short of the institution, science. A civil society whose level of education is higher than ever before demands a voice in the decisions on scientific-technological developments that touch on the self-understanding and value structure of modern societies. Even if by far not everything that scientific curiosity pursues is controversial, we cannot overlook the fact that the most promising and future-oriented areas of research, like the bio- and nanotechnologies and developments in biomedicine and the neurosciences, become the focus of public protests and rejection. At the center are questions of identity and protection of the private sphere, changes in kinship relations due to progress in reproductive medicine, involuntary exposure to risks, and the approach to the resources and services of nature. Many publicly expressed worries are based on the fear of losing control over how one leads one's life and on the threatening loss of the self in a confusingly complex world shaped by science and technology and in the midst of an inexorable process of globalization.

In one case, science is accused of no longer paying enough attention to the public interest and of becoming too dependent on markets and their economic constraints. In the second case, the accusation is that science is not public enough because it does not adequately take account of the legitimate claims of civil society and becomes too dependent on the state and industry. In one case, scientific curiosity is to be tamed by subjecting it to the regime of private economic use and its efficiency, while in the other case the domestication is to be achieved through a democratization of scientific expertise, including a public voice in the setting of scientific priorities. But these two directions are only seemingly opposites and above all should not be seen in isolation. The increase in prosperity in the Western industrial societies and the spread of the new information and communication technologies have led to a striking departure from the idea of a paternalistic and centralized welfare state that knows and can satisfy its citizens' needs. Instead, science increasingly counts on private and privatized means. The rhetoric of the empowerment of the individual, who best knows her own interests and knows how to and wants to decide on her own, merely underscores the attractiveness that private property and the power of disposition have won in the thicket of growing interdependencies and diverse dependencies—the promise of greater individual autonomy.

The democratization of scientific expertise, which is skeptical about the credibility of scientific impartiality and its disinterest, also demands that science provide greater public accounting and that research orient itself more toward what moves people positively and negatively today. The shift from the state to the market and the continuing privatization of areas that used to be within the purview of the state have created the figure of the freely choosing, freely deciding consumer and voter. It therefore also lies within the latter's competency to buy or reject the products coming to market that are owed to science and technology and also to judge other results of research and research's orientation in the future. Privatization thus not only is a powerful theme in neoliberal ideology and political rhetoric but also has captured the public imagination by promising greater individual autonomy. The freely choosing consumer is the twin sister of the authentic, free individual.

In this way, the two directions converge in the attempts to tame scientific curiosity. Privatization and "propertization" of the production and products of scientific knowledge are nothing else than the expansion of a regime that helped the industrial societies achieve their high degree of economic growth. Since science and technology are today regarded as the crucial driving forces for further growth and improvement of the standard of living, they too should be brought into the regime that has been successful thus far. The efficiency of markets, competition, and intellectual property rights should also prove its effect on the growth of productivity and the increase of scientific-technological output. A knowledge-based society also increases its production of epistemic things, various kinds of abstract objects, and technical artifacts that are subject to the same rules.

The democratization of scientific expertise is also merely the expansion of the principles of governance that have served the Western liberal democracies well. Today, science and technology are no longer viewed with awe but are part of everyday life. Mediated by the educational system and the qualifications and certificates people acquire, they determine people's chances for upward social mobility, their working world, and the course of their biographies. It is thus logical to extend the concept of citizenship to science and technology. "Scientific citizenship" comprises rights and duties and asks about both the functions that an expanded concept of citizenship could fulfill in social integration and also the duties that arise from it for citizens as well as for political institutions and administration.

The decisive question, of course, will be how far the attempts to domesticate scientific curiosity can go without endangering autonomy, which science will continue to require in the future. The autonomy of science developed out of currents and struggles similar to those that led to the concept of the autonomous individual. The freedom of action that science enjoyed was always also de facto limited and historically changeable. If the public character of science and its service to the public interest are under discussion today, this is also a result of the withdrawal of the state, which permits science as an institution to step out from its shadow and protection. It is more intensely exposed to the forces of the market but also to the demands raised and protests staged in the name of democracy. How far the vision of a privatized production of knowledge can go will also be measured by the degree to which it is politically acceptable. More efforts will be required to create public spaces for negotiating what scientific curiosity can and may do.[10]

Curiosity Receives Support: The Role of the Symbolic Technologies

Curiosity is a cognitive ability that the brain uses to explore the environment. To unfold curiosity's potential, the use of cognitive tools—particularly thinking, the capacity for abstraction, and the technical skills needed to produce material tools that change the environment—has to be embedded in cultural practices and anchored in a social structure. The human brain and its capacities are unique, not so much because of their biological development (which is not unique) but because of the human capacity to create and assimilate culture and pass it on to the next generation. The human brain and its capacities are the hybrid product of biology and culture. By itself, the brain can achieve little. The sources of experience may be initially individual, but for experience to be usable, it must be processed by culture and the synergies that result from interactions among many other human brains.

Paleontologists like André Leroi-Gourhan have long been interested in the close connections between hand and tool, face and language, and the influence of the motor functions of the hand and face on the connection between thinking and the instruments of material activity and sound symbols. When graphic symbols emerged, the formation of sounds and signs (graphism) was originally closely connected and only later replaced by a separation between the idea of the picture (art) and script, but today they are experiencing a renewed mutual rapprochement. The tool, says Leroi-Gourhan, "leaves the human hand early and becomes the machine: in the end, thanks to technological development, the spoken language and visual perception are subjected to the same process. The language that humans had objectified in the works of their hands, in art and script, now reaches the highest degree of its separation from them and they entrust their innermost phonetic and visual qualities to wax, film, and the magnetic tape."[11]

The rise and spread of the new information and communication technologies turned the development that was just beginning in Leroi-Gourhan's time into one of the key technologies of the closing twentieth century. Attention usually focuses thereby either on the economic effects, such as the increase in productivity, or on the scientific-technological achievements that lead to further advances, which were made possible by enormously increased computer capacities. Much less often are questions asked about the effects that these technologies, like other cognitive tools, have on the development of the human brain and the meaning of innovations in this context.

Along with language and the development of cultures of oral transmission, one of the most significant innovations in human history was the invention of script. The evolutionary psychologist Merlin Donald posits that the abilities to use script and symbols changed the functional organization of the brain. His hypothesis is that the use of script effects a cognitive reorganization of the brain individually as well as collectively, especially when the majority of the members of a society uses it.[12] Long years of schooling, for instance, enable people to achieve an adequate level of literacy (and thus of the use of symbols) in various fields—technical, mathematical, scientific, but also musical—and by mobilizing uncounted neural resources alters the way people think and carry out their work. Recent investigations of dyslexic Chinese and European children, incidentally, shows that, in all of them, a certain part of the brain functioned less well than in a control group of children without weakness in reading and spelling. But it also turns out that the regions of the brain activated when reading Chinese are different from those activated when reading Latin script.

Literacy is not the result of a Darwinian evolution. The biological evolution of humans unfolded long before they invented symbols. Literacy is neither natural and given nor universally distributed. Humanity existed for thousands of years without script, and most languages developed at a time when there was no form of writing yet. Nevertheless, all children can learn to read and write if they have the opportunity. The neuronal basis of literacy is cultural.

The ability to read and write is a consequence of the invention of external symbols—that is, their materialization and material depiction. But their effect does not end with the suspected reorganization of the human brain. Donald assumes that they lead to much greater changes, expressed in the whole of perception and recognition and thus also in how a human society thinks and how it remembers. New forms of mental representation arise. So these are extremely powerful technologies that work with symbols. Once invented, they unleash their inherent creative force and continue their effects on their own. From musical and mathematical notation systems to the broadest spectrum of artistic forms of expression, from diagrams through maps to multimedia imaging techniques, they comprise the entire spectrum of the material culture of our days. Without symbolic technologies, our society's scientific, technological, and cultural institutions and its highly technologized work achievements would be unimaginable. Symbolic technologies made it possible to build up an externalized cultural storage system that is available as a constant group memory and that, analogously to the individual memory, constantly changes and is active. The external symbols are themselves highly developed technologies. They are *cognitive machines* that change how we see, think, grasp, and deal with the world. Their central importance is that they free human consciousness from the limitations laid on us by biological memory. From an evolutionary standpoint, this is a radical innovation.

Whereas the preliterate cultures essentially had two technologies of memory storage at their disposal, storytelling and mimesis, today people have an enormously expanded set of symbol-technological instruments of preservation. Nonetheless, it is astounding how slow the development was from Ice Age cave art through the first calendar notations and earliest navigational aids to the invention of script and how long it took before the latter was reflectively used. For example, it is known that, in ancient Greece, the written notations of historical events made it possible for the first time to compare the accuracy of oral tradition with them. New concepts arose—like evidence and standards of validity and procedures for verifying them. When modern science began institutionalizing itself in the seventeenth century, it presented the demonstration of its experiments to the public, which functioned as a witness for what was seen and shown. But to this day, it has entrusted the real validation and certification of its results to writing, after appropriate quality control through peers. Publication in a specialized scientific journal aims not only to make the results public but also to put them into script.

Even if symbols are often invented by individuals, they build on the common stock of knowledge, and their use, their spread, and their further development is a collective enterprise to an even greater degree. To make full and efficient use of the power of symbolic technologies, a society must have corresponding tools, infrastructures, work habits, and communication mechanisms at its disposal. The presence of symbolic technologies alone does not suffice to trigger a cognitive revolution, and under certain circumstances it can even prevent one.* What was emerging at the beginning of modern science—namely, that cultural and social mixing and the reduction of class boundaries and hierarchies promote the creative flow of ideas and are the indispensable precondition for the circulation of knowledge—is also true for the use and societal spread of symbolic technologies.

For Merlin Donald, human creativity unfolds at the interface between the cognitive activities of the brain and the materiality of symbolic technologies—in art as well as in mathematics, the sciences, and the development of institutions. The power and efficiency of symbolic technologies is based on the existence of an externalized field of memory and operation. This permits mental operations at the disposal of consciousness to extend their radius of effect. Thoughts can be "externalized" by shifting them into an external field of memory and operation. This produces the necessary capacity for distancing, which is the precondition for every form of reflexivity. Games between the internal and the external can be invented, and experiments can be conducted between proximity and distance in which

* In history, for example, the tendency to secrecy has repeatedly prevented the spread of existing knowledge.

various viewpoints and arguments are tried out and one can practice putting oneself in the place of the other. Moreover, these fields of memory can be reformatted and changed in accordance with the intended use. Human consciousness acquires a mirror world that makes it possible to swing back and forth between internal and external representation. The capacity for a *multifocal attention* arises that has practice in mediating between the tensions resulting from the dynamic diversity of standpoints and perspectives.

Today, symbolic technologies form a cognitive and materialized network that is cast over society and that organizes itself and strengthens the production of knowledge and the emergence of the new. Their initially slow spread was followed by a phase of acceleration that is still far from reaching its limits. The rise and development of symbolic technologies underscores the entanglement between the biological equipment of the human brain and the cultural inventions that it has brought forth, whereby the social structures take the active and necessary role of coupling the two. This hybrid nature contributes to the emergence of the new and corresponds to the dependency of the unfolding of biological potential on culturally and socially produced conditions. It also underscores the double shape of every innovation. On the one side stand individual creativity and the uniqueness of the individual. The creative abilities and their effect can be retrospectively described and analyzed, but the decisive creative moment eludes observation. On the other side are the diverse interdependencies, the plethora of social and cultural dependencies on others and on a community that can foster or inhibit individual creativity. This interplay, which constitutes the dynamic of the new, has not been adequately studied since it was long overshadowed by the fascination of a concept of the genius that arose in the Renaissance and that lives today in the cult of the star.

A glance back at the often neglected premodern technological history of Europe reveals the degree to which precisely technological innovations depend on social and cultural factors, which can be joined by local and temporal shifts and simultaneities or nonsimultaneities. The technological historian Steven R. Epstein posits that the industrialization of Europe in the eighteenth and nineteenth centuries was the result of a long-term process of small but cumulative innovations that can be traced back as far as the Middle Ages. Although technological progress was slow in premodern Europe between the thirteenth and eighteenth centuries, it was lasting and proceeded without interruption. Flourishing periods were not interrupted by such long-lasting stagnation as they were in the great Asian cultures of China and India. In addition, the geographical centers that led in technological innovations increasingly shifted from the south to the northwest. In each new region that they penetrated, the innovations mixed with the local given situation and found a way to combine with it to produce further steps of progress. Such a mixing process of technological diffusion and recombination under different social, economic, and institutional conditions was completely lacking in premodern Asia. Third, says Epstein, all premodern technological knowledge—knowledge of how one makes things so that they function—was conveyed by persons. Technological knowledge is embodied in its practices. Geographical shifts or the rise of new leading centers is possible only if those with practical knowledge take it with them. This kind of mobility was more possible in Europe than in Asia because family bonds were less rigid and locally anchored and because, in Europe's fragmented political and economic systems, a competition in tendering bids could promote mobility.[13]

At the beginning of the twenty-first century, premodern technology can still be encountered locally, but "knowledge of how one makes things so that the function" has reached a highly technologized and science-based level. The contrast between nature and culture—between, on the one hand, the natural and the phenomena, organisms, and components occurring in nature and, on the other hand, artificial products created by people—appears to be dissolving and merging in a "vireality," a fusion of virtuality and reality. But we have not yet arrived in cyberland, despite the fascinating and frightening glimpses that, warning or enthusing, reach us from there. Reports so far see humanity's future home as an inhospitable place.[14] Before such a future draws us into its thrall and brings about this imagined fusion of brain activity and emotions, of bodies and technology, that the new biotechnologies and other convergent technologies have brought at least into the realm of possibility, we should ask again how far curiosity can extend into the future and what answers, if any, it can find there.

The Curiosity to Know the Future: Nature Knows No Future Tense

When the introduction of the new is controversial, when resistance appears and opposition forms, invoking a body of higher, morally bound values, the issue is often unsettled questions of placing a boundary between nature and society. This dichotomy may be false and constructed, empirically and historically disprovable, but it persists with a stubbornness that is again and again mustered to defend ingrained interests. In the conventional argumentation, nature stands for the given and for something willed by a higher power. Nature is regarded as the norm that defines normality. Human intervention, by contrast, stands for a transgression of set limits that subversively undermines and endangers normality.* But the attractiveness of invoking a nature that stands for the immutability of normative claims and for stretches of time that exceed human scale does not take us far since the next step requires us to decide where the domain of nature ends and who is authorized to determine this. The arduous debates and struggle over a societal consensus will increase in the future.

The belief in the immutability of nature screens out knowledge of the societal processes that lead to knowledge and the capacity to intervene in, manipulate, and control nature. Resistance against the new is often rallied in the name of an authority that seems to stand outside of societal norm-setting and that is therefore granted a right to set pre- or suprasocietal norms. At the beginning of the modern era, when there arose for the first time in Europe the modern concept that politics is neither guided nor secured by God's pleasure, the focus was on the problem of the practice of human freedom and the limitations

* Biologists also make this distinction, speaking of changes caused by humans as supernormality.

to be placed on arbitrary political power. Two institutions came into question because of this focus. One was the legal system, which was to be primarily responsible for working out the norms, regulations, procedural directives, and guarantees that gradually determined the construction of the constitutional state under the rule of law and for ensuring their implementation. The other institution was science, which was equally young and hardly institutionalized yet. According to Yaron Ezrahi, an Israeli political scientist, science was historically conceived as an apolitical authority with the capacity to discipline political activity, criticize decisions, and place limits on the secular state. This—political—function of science was sometimes used to depoliticize and veil the exercise of power through the invocation of scientific and technological rationality. But the certification of a science-based reality contributed crucially to replacing the will of God with the laws of nature and led to a restraining of the arbitrariness of political action.[15]

Even in its beginnings, modern natural science showed that mechanisms could be found with which the scientific community could reach consensus on disputed issues. In a time wracked by religious and civil wars, this doubtless contributed to solidifying the moral prestige of this young institution, which nevertheless was wise enough to accept the limitation placed on it by the state and religious powers—namely, not to interfere in their affairs.

The authority of science as an apolitical authority is based on a legitimacy derived from the authority of nature. The natural laws it investigates are higher laws that are nonnegotiable and cannot be subjected to a state power or a judge's ruling. Science certifies that the reality of nature is removed from the jurisdiction of human laws and their arbitrariness but that there are nonetheless procedures "to reveal its secrets," to manipulate it and make it serviceable. This makes the institution, science, the apparently infallible mediator between nature and society. Science claims the authority to speak in the name of nature. To successfully advance its program of exploration, it must be free of state and religious interventions for it "speaks truth to power." The other way around, through its procedures of quality control, especially peer review, the scientific community guarantees that the knowledge produced is reliable. These two strands—science's claim to speak in the name of a higher, apolitical authority and its claim to guarantee to the public the reliability of the knowledge it creates and certifies—are the basis for its autonomy to this day, even as the public character of science is coming under pressure.

In this way, science became the referee in all questions that it can answer on the appeal to nature and its own apolitical authority. But the more successful and consequential human interventions in nature became, the more clearly visible were the limitations of this institutional arrangement that was otherwise so successful for science. The faster the pace of scientific-technological innovation, the greater the proportion of "social knowledge," which, in analogy to Epstein's definition of technological knowledge, I would like to define as knowledge that knows social contexts and ensures that they function. Scientific-technological innovations have to be emplaced in already existing organizational forms, social structures, and biographies. To be successful, they must be accepted and altered in such a way that they identify and meet latently present needs. They must contain answers to only diffusely articulated expectations, and their promises must be redeemable not only as "technological fixes" but also within the disorder of social reality. The rapid path via technological solutions, as practically effective as it can sometimes be, runs the danger of overlooking social contexts that cause the problem and that can reappear all the more insistently in another place.

In the public discussions of controversial questions related to values and ethical principles, it is also becoming ever clearer that nature—and therefore also natural science—proffers no standards for human activity, so science cannot answer these questions clearly. Thus, Hubert Markl, then the president of the Max Planck Society, wrote that the German debate on stem cells, which is probably conducted more heatedly in Germany than anywhere else, entails far more than the ostensible topic. It refers not only to stem-cell research and preimplantation diagnostics but also to the questions of what it means to be a person and what the freedom and tasks of science are. To be a person, Markl argues, is a culturally defined attribute rather than a biologically determined fact. Even if persons are biological beings, being a person goes beyond that. This is why a person cannot be defined solely in terms of molecular genetic facts, like the 43.2 million nucleotides that are arranged in a specific way inside a zygote. Personhood is thus also defined by the culture, which gives us many different answers.[16]

Science never completely managed to prevail in enforcing its monopoly on interpreting an "objectively" graspable reality or in reducing the content of public controversies to "hard facts." On the one hand, publicly discussed values stubbornly resist being cleanly separated from the supposed facts since on closer examination the facts often turn out to be the results of determinable or preset political and social framework conditions or questions. On the other hand, the concept of scientific objectivity is itself subject to processes of historical change, as the history of science shows. To maintain its claim, the processes for obtaining objectivity would have to adjust to the changed technological givens as well as to the changed social dynamics of the scientific community.[17] In addition, alternative claims to interpret reality that have withdrawn from science (for example, alternative medicine) were able to survive in premodern niches, where they fulfill other social functions.

How strong the alliance between science and the state actually was went unnoticed until it began to break down. What welded them together in the twentieth century was the systematic use of science in the two world wars, followed by nuclear arming and other forms of military research in the time of the cold war. This influence of the state on science and the state's claim to the right to use its monopoly on knowledge paradoxically became especially noticeable in basic research. Basic research produces decisive breakthroughs and new scientific knowledge but often at the same time also new technological applications. The freedom of action needed for theoretical and technological-practical basic research was granted, but then the new knowledge and technological artifacts were selected for military-operational utility. With the end of the cold war, this regime of close alliance was replaced in part by a decentralized form of funding research that is oriented toward worldwide economic competition.

With the advance of the market and transatlantic economic interdependence, the nation state lost further ground and importance. On the one hand, the assumption that scientific and technological standards of rationality guide state policy was also undermined, and Enlightenment thinking was replaced by so-called postmodern configurations. The media take a significantly larger role in democracy mediating the public realm and thereby in redefining the boundary between public and private. The institutional constraints that separate, including normatively, the realm of politics from the economy and the private sphere are mixing in an often unclear way. At universities in the United States (but not only there), new guidelines are repeatedly promulgated whose function is to harmonize academic freedom with industry's interests in fostering research. Specialized journals, especially but not exclusively in the life sciences, demand that submitters reveal possible conflicts of interest, and ethics commissions are supposed to ensure that commercial interests do not hinder research's interest in publishing negative as well as positive scientific results.

If today mistrust, tension, and loss of credibility in the relationship between science and the civil society are lamented in many places, the reasons are in part to be sought in the dwindling normative consensual basis in liberal democracies that oriented themselves toward scientific rationality and the technocratic structures of the state and that were supported by science's apolitical, neutral stance. While the state relinquishes some of the competencies it used to exercise to marketlike institutions in the name of greater economic efficiency, the institution, science, has to make greater efforts to credibly maintain its apolitical neutrality in the public sphere. Scientific and technological expertise is in greater demand than ever before, but the reason for this is the increasing density of regulation and the growing complexity of decisions with a scientific-technological content. The greater demand is accompanied by the equally greater vulnerability to objections and protests made in the name of a pluralistic political diversity.

The current self-understanding of the pluralistically organized liberal democracies prevents any one authority from having the right and the credibility to legitimize decisions or in general to speak in the name of nature, which stands above society. Society's share in the coevolution of science, technology, and society has grown too strong to permit such a solution of conflicts. The twentieth century, which was one of devastating, frightening political projects as well as of monumental scientific-technological projects, assumed in its megalomaniacal optimism about progress that the guardrails and coarse patterns of orientation that modernity had normatively set up for societal development and progress could be followed straight to the future. Today's thinking is turning away from this belief. The advances toward a scientific-technological future are surrounded by imponderabilities and uncertainties. Politically, this is expressed by politics' fragmentation as the reverse side of the coin of pluralism and by its short-term thinking. Science and technology have lost some of their privileges and part of the protected space that the state and politics granted them. Above all, they can no longer invoke higher criteria of rationality to offer those securities on which the human capacity for judgment, decisions, and action could be anchored and legitimated. The rationality of action has been splintered into a diversity of rationalities that invoke an explosive and unstable mixture of economic and private interests, scientific and technological orientation, remaining state competencies, and the demands of a civil society—all this in a public theater that the media illuminate and stage.

A normative basic consensus can be found, at best, by accepting an openness toward the future and by viewing society as a laboratory serving the public good—under limitations that must be concretely worked out. Instead of being able to trust in secured prognoses for the future and in a science that acts mostly apolitically and independently of external interests, what is needed are pluralistic negotiation processes. Nature provides no answers for most of the questions that arise from the growing knowledge of its functioning and from the need for decisions and action that results from the successful interventions in and manipulations of it. This means that we must accept effort and conflicts and work to create public spaces in which we can negotiate what cannot be otherwise decided.

But this does not mean that nature has disappeared as a normative authority in societal discourse. In this regard, it is useful to distinguish between images of science existing in society from images of nature in societal discourse. The images of science are fed from a broad spectrum of sources of imagination, including those that science projects onto society. As the historian of science Simon Schaffer has shown, they oscillate between two poles. At one end, scientific activity is presented to the public in a way that suggests its direct connection to everyday experience and everyday understanding. Scientific activity then becomes a continuation of general thinking, and all people can understand it. The (even today still usually male) scientist acting in this image resembles a hobby tinkerer. He has craftsmanlike skills like those encountered in everyday life and is inventive and practical.

The opposite pole is represented by an image characterized by an unbridgeable separation from everyday experience and everyday understanding. Here, science appears as an extraordinary activity open to only a few especially gifted minds and leading into realms that unmistakably go far beyond everyday understanding. The accompanying image of the scientist (again usually male) is of someone passionately and totally living his ideas, someone who is guided exclusively by his curiosity. This image of science finds its counterpart in the idealization of basic research, which stands sharply apart from expectations of utility. Schaffer says the two images are evoked alternately as circumstances demand. The heroic figure with the characteristics of genius who is above all the economic and other constraints, preconditions, and consequences of her action lives unproblematically alongside the practical figure who is found everywhere in everyday life.[18]

The images of nature that are present in societal discourses also oscillate between two opposites. On the one hand, nature is imagined as the epitome of threat. Its powers can be tamed to a certain degree by means of human cleverness, knowledge, and technological imagination, but this state is never lastingly guaranteed. Since nature will always be more powerful than people, it is unpredictable. Anthropomorphically speaking, it can "strike back" at any time and even "avenge" itself for what has been done to it. On the other hand, an image of a nature requiring protection is also imagined—endangered species, diminishing biodiversity, continued degradation of the environment and its resources, which all must be brought to a stop. In one case, nature becomes a deterring example that, deeply amoral, comprises every imaginable brutality and cruelty. In the other case, it becomes a moral role model that can teach us how to preserve diversity and how to nurture it. In both cases, nature is normatively charged and normatively interpreted. It becomes a mirror of cultural achievements and equally of their failure, a mirror of ourselves that reveals deep-seated ambivalence. But the gaze is unable to see itself.

The success and the failure of freedom are both inherent in the project of modernity. A glance back into the past provides adequate evidence for both options and permits at least a careful weighing. Clinging to utopian and dystopian ideas, wishes, and fears darkens the glance into the future. In the face of the new, the play of imagination can begin, but communicating the new is already difficult because the language for this is lacking and the images are deceptive. Different patterns of argumentation can promote or hinder the new. The language that the new is clothed in is either not yet available at all or not yet made for it. It triggers varying associations and holds much scope for interpretation. It easily falters where the point is to dampen the shock of the new, which nonetheless has to struggle on its radical path to a changed way of viewing things. A historical example can illustrate this.

For almost no other radical theoretical edifice of ideas is this as true as for Charles Darwin's *The Origin of Species*, published in 1859. Darwin was very aware of the problem. When Wallace's surprising letter from the South Seas reporting on his notes and theoretical considerations, which mostly corresponded to Darwin's own, reached Darwin one day, Hooker, Lyell, and other friends of Darwin urged him to begin compiling the observations he had painstakingly noted more than twenty years earlier along with the associated theory (which he originally termed a mere "abstract"). Darwin clearly understood that one of his biggest problems thereby would be to clothe the knowledge he was introducing in a new language that would enable its acceptance and stabilization. Whereas other natural philosophers of his time consciously relied on predictions, experiments, demonstrations, or mathematical proofs because they thought they saw in them the means for public accreditation of scientific truths, Darwin's approach was doubly unusual.[19]

His content demanded a lot from his readers. He spread before their imaginations a world ruled by irregular and unforeseeable contingencies. He revealed a nature that was cruel and full of errors and in which there were no moral laws or purposes. Animals and plants were not the result of a creator's special design or plan. As he wrote at the beginning of his primary work, Darwin was firmly convinced that the species were not immutable. He asked his readers to accept his answers for the sole and simple reason that they were the facts. To this purpose, he invited them to participate in the information, observations, and sorted evidence he had gathered over a period of more than twenty years and to accompany him in carefully weighing the conclusions he had drawn. In the text, many different strands of narrative and argumentation are interwoven and enriched with rich, inventive metaphors. But ultimately what counts is the structure of the arguments Darwin mustered to persuade his readership.

Although Darwin presents the sequence of chapters as if they corresponded to the chronology of his daily research work (thereby implying that his ideas developed from the observations and facts), the actual course of the development of his theory was much more complicated and many-layered and was certainly not linear. The concept of "natural selection" was not simply present in nature, nor could it be presented under the injunction to "look and understand." As Janet Browne underscores, Darwin had no mathematical equations or formulas in hand to prove his theory. His book contains only words borne by strong convictions, visual presentations, the weighing of probabilities, interactions between large numbers of organisms, and the subtle consequences of tiny coincidences and changes. He often employed analogies between what was known and what was unknown. He relied on statistical frequencies to support his arguments. He consciously employed a strategy of progressively weakening his reader's resistance by adducing factual examples. And of course he was aware that the whole of his factual material was inseparable from his theory. In a sentence that has become famous, he expresses this in his characteristically simple and disarming manner: "How odd it is that anyone should not see that all observations must be made for or against some view if it is to be of any service!"[20]

Literary historians and historians of science have often commented on Darwin's skillful use of language and his cultivated, accommodating manner, which won readers' sympathies. It can be shown that some of Darwin's arguments that appeal to nature have the purpose of giving him the necessary cover for the novelty of his ideas. Gillian Beer investigated the question of why Darwin uses the term "natural selection" rather than simply speaking of "selection."[21] "Natural selection" presupposes a counterterm, "artificial" selection. The term "natural selection" is based on the analogy with "artificial selection" and thus on an analogy between the processes of selection, which must unfold either under natural or artificial conditions. Artificial selection includes all human activities aimed at effecting rapid genetic change—in agriculture or in breeding pigeons, dogs, or orchids. It is based on a technology, accessible also to amateurs, that includes human aims as well as intentional results. It includes the market.

Gillian Beer suspects that Darwin made this distinction not least because of his experiences on his journeys on the *Beagle*. On these journeys, he had seen the devastating effects of colonial conquest and the genocide committed against aboriginal peoples. In *The Origin of Species*, Darwin avoids mentioning humans, knowing full well that he would trigger controversies if he did (he speaks of his own "abominable volume"). Gillian Beer concludes that the term "natural selection" allows Darwin to pursue the questions that the term "artificial selection" excludes—questions of nature's long temporal scale in contrast to the short-term scale of breeders and of nature's complete indifference in contrast to human intentions and their drive for mastery, planning, and management.

Darwin's theory distinguishes itself clearly from the views of his contemporaries (with the big exception of Alfred Russell Wallace)* and from the preevolutionary thinkers. A key to this is that for Darwin diversity and deviation are central principles of survival. The truly creative forms, the decisive difference, are owed to the nonconformity that arises from being different. For Darwin's theory and its reception, that only increases the urgency of the problem that the "natural" is frequently equated with what is considered normative. For Darwin, deviation, rather than conformity, is decisive for survival. Differences from parents, rather than similarities to them, are what count because this is the only way that existing ecological niches can be expanded and the exploration of new niches can be nudged

* In 1858 on the South Sea island of Halmahera, Wallace wrote his famous letter in the form of an essay, titled "On the Tendency of Varieties to Depart Indefinitely from the Original Type," to Darwin, triggering confusion and despair in the latter. Darwin's friends Lyell and Hooker thereafter arranged a meeting of the Linnean Society at which were read aloud a previously unpublished essay that Darwin wrote in 1844 and the abstract of a letter that he had written to Asa Grey the year before. Both contain the Darwinian theory of natural selection and the origin of species through modification. Thus, Wallace's essay came later, and Darwin's priority was ensured. Darwin's book *The Origin of Species*, whose composition was triggered by Wallace's essay, was published soon after. Darwin became famous, while Wallace was forgotten for a long time.

forward. Nature always moves on the verge of the monstrous—and according to the laws of evolution, it must. For what may initially seem like a monster can later prove to be a new type that has merely waited for the proper conditions before developing. At the right moment, what was initially monstrous can become the norm and define normality.

Whereas the concept of natural selection expands, the concept of artificial selection narrows. If the first stands for exuberance and plenty, the latter indicates parsimony and the ruthless exclusion of everything that is not adapted to current conditions. In a small population, selection does not mean fertility but extinction. The intrusion of an alien species can lead not to the native species' improvement but to its disappearance, as Darwin was able to observe in the result of the encounter between the Spanish and the *Indigenes* in South America. He had seen how the invasion of the Spaniards made the aboriginal population regress from settled to nomadic, quite in contrast to the stepladder of the "natural" rise of every civilization.

Darwin was thus able to denude one of the nineteenth century's most prevalent terms of its "naturalness." It is usually worthwhile to question the self-evidence of terms and concepts that are employed. But it is seldom so important to make oneself aware of their function as when nature and what is regarded as nature's binding authority are brought into play as arguments. Nature stands precisely *not* for immutability but for diversity and deviation from what has existed heretofore. If everything is constantly changing, then "natural" can be understood only as an appeal to the knowledge of the past. Then it is an expression of the attempt to ward off the future and to ensure that deviance is forgotten. In the following, Gillian Beer summarizes the role that "nature" has to play in argumentation:

> Nature is how things *have always already been*, i.e., lasting and universal.
>
> Nature is how things were *in the past*, i.e., ideal and nostalgic.
>
> Nature is how things *normally* are, i.e., as a matter of course and the norm.
>
> Nature is how things *should be* or *should have been*, i.e., as described by imperatives and ideals.
>
> But if we want to know how things *will be* (instead of how they should be), this is not part of the concept of "nature."[22]

Nature knows no future tense—and yet it constantly provides for the emergence of the new. What begins with insatiable curiosity, the exploration of the unknown, and the inexorable appropriation of a future that nevertheless eludes all claims to possession—the emergence of the new—follows differing paths. Nature has long time scales and the laws of evolution. Human societies, which themselves have emerged through evolution, also have cultural and social forms at their disposal. They are able to imagine their future and various designs for it. The capacity to aspire helps these imaginings to come to realization, although under manifold limitations. As fragile as the future may appear and as fragile as it is, conceiving it anew and conceiving it as different grows out of what has already been achieved. The unprecedented potential of scientific knowledge and of technological abilities opens up a tremendous realm of possibilities. The collective wager on the future that we have made goes by the name of *innovation*. But it, too, is unable to say how things will be or how they should be. To this end, societal debates are needed, along with strategic sites for finding a consensus. Language is needed to describe the new, and selective appropriation is also required. Needed are the cultural resource of reflection and the attitude that has accompanied us since the beginning of modernity, ambivalence. It contains a yes and a no—and nonetheless permits us to act in the face of a fragile future.

2 Paths of Curiosity

The question is the source of fertility, and its lack is one of the reasons for sterility. . . . The question suspends the given meaning of the world: it is a work of discovering, of taking apart, and of releasing.

—Marc-Alain Ouaknin, *Bibliothérapie: Lire, c'est guérir* (1994) (commentary on Genesis 15:1–4)

The Absurd Bifurcation of Nature

When we speak of curiosity, we immediately think of the obstacles in its way. For curiosity, even if we grant that it is the driving force that can take us beyond what is already given, sees itself confronted again and again with limits we place on it—bans on thinking, fears, and warnings to be cautious—as if it were necessary once and for all to dissolve, smooth, and eliminate all the contradictory impulses it feeds on.

Measured against earlier promises and hopes, it may have brought us far—whoever may be concealed behind the collective *us*. And yet as modernity's promises are successfully fulfilled, their shadows grow. The stakes of the game seem to increase with every success that is celebrated. Today science and technology intervene overtly and covertly, directly and indirectly in cultural self-understanding and alter our perception and perhaps soon our emotions.

Aren't we approaching, faster than we could ever have imagined, the time when the human is inseparably mixed with things made by humans, a cyborg epoch that has often been imagined but never realized? A time when we must finally lose or relinquish control over ourselves, now, when we finally believed we had achieved it?

The fear of the new inhibits curiosity, even if curiosity is not completely intimidated by the fear it creates. But curiosity begins to hesitate and wobble. It seeks escapes and wants to insure itself. And so the strategic considerations begin. The game of possibilities becomes a doubled game or turns into games that are played in many places and whose point is to bring forth their own rules, just as evolution produced its own rules, including rules governing how the rules change. Curiosity, insatiable as it is, thereby drives us forward. It seeks new paths and willingly accepts that some are wrong turns. It seeks risk, thereby repeatedly staking what it has already found and achieved. Again and again, it subverts limits that have been reached to fence them in or guide them in certain directions. It poses questions that are not permitted, and unwise as it is, it presses for action even where it should draw back. Questions instruct it and always point beyond what is given. It speculates, tries out, and has great difficulty learning that it should consider the consequences. It can be contagious, but it can also be smothered. It is currently attempting boundary crossings in the grand style, without making a great ado but persistently and with its intrinsic matter-of-courseness. It asks about the quality of the boundary between nature and nature—namely, between "external" nature and the nature supposedly internal to the human being—and about how much more permeable, indeterminate, and mixed the boundaries between nature and society can be made. What drives it is the pleasure in playing and the impulse to move beyond what already exists. It is fundamentally irresponsible because it is obligated only to itself. But it is a part of us and of our nature. What we do with it, what we make it into, and how we deal with it determine how we live together. It sets the limits we place on ourselves—only to transgress them anew in the future.

The mathematician and philosopher Alfred North Whitehead saw one of the limits blocking curiosity in the bifurcation of nature into two different systems of reality. One system is an objective part that can be mathematically formalized and empirically verified. This reality consists of units like electrons. It is a reality that is there to be known. But the knowledge we have of it is another reality created by our brain and our senses. Nature is split into two parts—a nature of perception and a nature consisting of the causes of perception. The subjective opposite pole comprises all meaning, beauty, significance, and values that we attribute to the objective reality. But such a separation (still deeply anchored in Western thought in philosophy and in the speechlessness of the disciplines) fails to understand that no fundamental difference exists between our access to knowledge and our ways of knowing.

Isabelle Stengers provides a new reading of Whitehead.[23] She locates the starting point of the "adventure" that Whitehead's speculative philosophy embarks on in the bifurcation of nature. This bifurcation is absurd because it sets up bans on thinking, dismisses the questioning of established concepts as inefficient or ridiculous, and precisely thereby contributes to the incoherence according to which nature is divided into two different and incompatible registers of speech. The bifurcation of nature stands for all the other dichotomies and divergences that make a decision necessary, a weighing of pros and cons, with the goal of taking a position. Instead, Whitehead tries to overcome the split and to set in motion a social transformation, similar to a healing process, that grants validity to subjective experience. He is confident—and here his experience as a mathematician comes into play—that solutions are possible to dissolve seemingly indissoluble contradictions if one is prepared to introduce new premises. As a mathematician, he creates solutions, but the creativity in finding the solution arises from the limitations imposed on the solution. The mathematician Whitehead was aware that mathematics finds solutions by inventing them.[24] The separation between experience and experiment, between the subjective and objective sides of reality, becomes dispensable for him.

Out of the absurd division of nature arises the incoherence that can escalate into the incompatibility of different registers of speech and ways of thinking. To be communicated, the new always requires language, but it often presents itself in two different languages that follow the pattern of bifurcation. For example, if the latest research on the neurosciences and the biotechnologies, which have the potential of altering the entire prior spectrum of human feelings, are presented to the public solely in the language of the neurosciences, then this provokes the other, subjective register of language to reply and contradict. The subjective language then speaks of the threat to human subjectivity or to freedom or to the illusion of this freedom. The incoherence could not be greater. While the one side celebrates freedom or its illusion, the other insists on the description of a world that is cold and indifferent toward this freedom or its illusion. This absurdity led Whitehead to radically question the existing concepts and terms. He wanted to renew them and modify them accordingly. To this purpose, he opened himself up to a "wild thinking" and a speculative philosophy that trusts that there are possible solutions.

For Bruno Latour, the original sin consists in the split between nature and society, which is also manifested in two different practices whose separation and distinction are the necessary preconditions for their efficient functioning. Modernity's original sin is its foundation, enabled its triumph, and lies in its hypocritical ambiguity. In *We Have Never Been Modern,* he accuses European modernity of always having spoken with a forked tongue and of having acted differently than it spoke. The one bundle of practices produces continual mixtures between various species of organisms. These practices create hybrids between nature and culture, both of which we belong to. The second practice, however, denies the first. Through a process of "purification," what was initially brought together is ontologically separated again. Sharply separated zones arise; the one zone contains human beings, while everything nonhuman belongs to the other zone. This ambiguity of European modernity is the source of its efficiency because the work of mixing (that is, of the creation of hybrid products between nature and culture) would be slowed without the work of "purification" (that is, of separation). At the same time, the work of mixing the human and the nonhuman, nature and artifacts created by humans (which is denied with the next breath), is the foundation for the great distinction between modernity and all the societies that have remained behind—the premodern societies.

Thus, according to Latour, modernity consisted of the separation between political power and the power that is based on scientific reason. In their deeds and their actual relationships with the others, the Europeans have always known how to support themselves with the rationality of violence and the violence of rationality. They remained undefeatable, while the subjugated were forced to become premodern. How could they have withstood and with what concepts and terms could they have put up resistance when these can always be recombined: terms like a *transcendent* or *immanent nature*; a society created by humans or a transcendent society removed from human influence; a distant or a personal God? Latour pleads to include the nonhuman actors—structures like neutrinos and crystals, viruses, tectonic plates, frogs, and other nonhuman things with which scientists interact and that quite obviously act themselves. In the parliament of things, Latour wants to give them a seat and a voice, and in the face of ecological crises, the division between nature and culture should be suspended since it was never actually carried out except in the conceptual work of purifying categories.[25]

There is no doubt that today there is a manmade proliferation of hybrid mixtures of all kinds that outdoes all the monsters, chimeras, and bizarre creatures that nature has thus far brought forth in the blind process of evolution. This continuing production of an artificially created nature, which itself in turn transforms itself into society and culture, brings together in creative and productive, destructive, or indifferent manner what nature never foresaw and what society need not have intended. It prepares the ground for further experiments and establishes the framework conditions for the people of tomorrow—how they will live and feel, how they will deal with each other, and how they will reflexively perceive themselves. A coevolutionary process is at work here, simultaneously blind and seeing, wanting to foresee the consequences and yet having to accept the unforeseeable. It is this *condition moderne* in which the absurdity of the split triggers so many emotions. For nature and its part in the technically and scientifically possible interventions remain morally neutral, even if moral sensibility may have evolved through evolution. Nature offers no answers to moral questions. It can at best mirror what we project into it and remind us that there is no inherent reason to prefer the natural and no reason to realize everything that is technologically feasible. The new battle zones opening up in the fields of bioethics are not really that new, at least in terms of the content of the conflicts. The explosive power inherent in them leads back to questions that were already being vehemently discussed in the 1930s and whose roots can be traced back to the Enlightenment.

Daedalus, Icarus, and the Science of the Future

Standing in front of the student club the Heretics in Cambridge in 1924, the young J.B.S. Haldane, who eventually became a famous biochemist, delivered a lecture later published under the title *Daedalus, or Science and the Future.* He was inspired equally by scientific visions and by the future socialistic society in which science would be the primary determinant of the progress of human civilization.[26] "Science," he wrote, "is the free activity of men's divine faculties of reason and imagination . . . it is man's gradual conquest, first of space and time, then of matter as such, then of his own body and those of other living beings, and finally the subjugation of the dark and evil elements in his own soul."[27] His youthful self-confidence, his comprehensive knowledge of the state of the sciences of his day (he was equally familiar with its art and literature), and his scientific and political enthusiasm tempted him to make far-reaching predictions about the contributions that science would make in the future to the solutions of humankind's urgent problems.

Haldane predicted the rise of biochemistry and the biosciences as a whole. Almost all inventions of practical utility have a physical-chemical basis, and many new inventions in the field of biology will build on them. Among them are contraceptives, the application of endocrinology, and the invention of many other substances that (like alcohol, tea, and tobacco) have stimulating or pleasant properties. He has a future student, "a rather stupid undergraduate member of this university," read a paper in which he summarizes for his academic adviser the most important scientific developments and societal problems. Among them are the triumph of the eugenic movement, the introduction of artificially created, nitrogen-fixing organisms, the artificial alteration of character by means of applied endocrinology, and finally "ectogenesis," the artificial development of the human embryo outside the womb from egg cells cultivated *in vitro* and artificially inseminated. It is clear to him that society, especially people who hold traditional values and ideas of morality, will not accept these achievements without objection.

Haldane was aware that he lived in a time between two world wars. The first had displayed the terrors of such a war, and after it many people nursed hopes for a world government ("it may take another world-war or two to convert the majority"): "The prospect of the next world-war has at least this satisfactory element. In the late war the most rabid nationalists were to be found well behind the front line. It will be brought home to all whom it may concern that war is a very dirty business."

The future, said Haldane, would not be a bed of roses, and the only thing that can be definitively said is that no doctrine, no values, and no institutions are safe from the progress of science. But he thought moral qualms should not be taken too seriously since traditional value systems and the distinction between good and evil are subject to change. Because religious traditions cling to unchanging values, there can be no truce between science and religion. Ultimately, society must adjust to scientific progress—or it will perish.

A prominent source promptly raised a serious objection to this unbroken and aggressively articulated belief in progress, in which humankind has no other option but to accept the positive achievements of science. Bertrand Russell, under the equally telling title *Icarus, or the Future of Science,* questioned the attractive image of the future in which scientific inventions would promote human happiness. He said it was more likely that science would be used to bolster the power of ruling groups than to make people happier. Once science taught it to fly, humankind's fate could mirror that of Icarus. But Russell's main argument aimed at science's failure so far to bring human passions under control to the degree that it brought nature under control. People are not rational; they are a bundle of passions and instincts. The progress of the sciences has changed the balance between instincts and environmental conditions, he wrote, and therefore the most urgent task was to tame people's rivalries and lust for power if "industrialism" was to triumph with the aid of science.

Russell sees clearly that progress does not come simply from the physical sciences but also requires social organization. The question, however, is not whether but how much and what kind of organization is needed. The "anthropological sciences," too, would have important contributions to make. Control over emotional life, using hormones and similar substances, would become more important than intelligence tests. The men who administer this system would have power exceeding the imagination of even the Jesuits. Russell arrives at the conclusion that science will permit power holders, in particular, to achieve their goals more completely than before. If these goals are good, this is a gain; if not, it is an evil. "Science is no substitute for virtue; the heart is as necessary for a good life as is the head." The heart stands for the totality of all friendly and generous impulses that ignore one's own interests. But so far, science has failed to give people more self-control, more friendliness, and more power to restrain their passions.[28]

The undeniable fascination exerted even today by the equally brilliant and virulent intellectual and scientific debates of the 1930s must not obscure the fact that the framework and the problem fields have meanwhile greatly changed, politically, economically, and culturally. First, the debate is no longer conducted within a relatively small, elite circle of participants. It is carried out publicly, contributed to and intensified by the mass media. It is pressured to hold science publicly accountable and to democratize scientific expertise. Today, no one any longer expects the sciences to be able to provide the rational foundations for the societal or political order. On the contrary, despite the sciences' spectacular successes, which fulfilled or exceeded what was once only imagined, the prestige and credibility of science have declined in the public realm. Even the assumption that a rational worldview would displace the importance of religion, replacing it with trust in science and technology, has proven false.* Yaron Ezrahi suspects that the current decline of belief in science's capacity to solve the problems of human civilization coincides with two events—the dissolution of the Communist bloc (which for Haldane and other prominent scientists long embodied the hope of productively uniting science and socialism) and a disenchantment with the attempts to bring together science and democracy (as John Dewey and others

* Pippa Norris and Ronald Inglehart, *Sacred and Secular: Religion and Politics Worldwide* (Cambridge: Cambridge University Press, 2004), point out that the economically highly developed industrial states in which secularization is most advanced show the lowest degree of trust in science and technology—far less than the trust displayed by countries with low levels of economic development.

promoted and hoped for). Both events indicate that there have been fundamental misunderstandings in the relationship between science and politics.

Science and technology have developed under both authoritarian and democratic regimes, albeit in different ways. But if, under diametrically opposed political preconditions, the expectations placed in science have historically failed, then we must reexamine the suppositions about the function of politics and its relation to science that underlie those expectations. The roots of the exaggerated expectations that Haldane and other contemporaries placed in the rationality of science and in its direct society-transforming power (which were still influential into the 1970s) reach back down to the critical countercurrents of the Enlightenment, especially to Rousseau. They are borne by the—false—assumption that people can ultimately agree on what is good and that political decisions on goals can emerge that can then be fulfilled by science and technology. This runs counter to the experience of Western liberal democracies that there are insoluble conflicts of values and interests. If no basic consensus can be achieved, then there can also be no rationally compelling decision among political alternatives. This means that the scientization of public life that the Enlightenment thinkers dreamed of is impossible.[29]

The absurdity of the bifurcation of nature into objective and subjective sides seems to have turned into an equally absurd bifurcation of society. Knowledge, even rational knowledge, cannot be automatically translated into action. Haldane's elitist assertion that society must adjust its values to what science offers not only founders on democratic objections but never really gets off the ground in the face of the regulating effect of the market, which selectively feeds back to what is taken up from the plethora of scientific ideas and technological possibilities. The harmless and endearing amateurs who bred pigeons or orchids in Darwin's times and thereby conducted "artificial" selection have been succeeded by professional innovation systems fueled by the market and the state. These consist of an impressive host of entrepreneurs, joint-venture funders, small start-up firms, stockbrokers specialized in blue-chip investments, technology-transfer bodies, and others who are assiduously bringing constantly new products and processes onto the market. And here we see a shift. Whereas the most important sites of technological change in the twentieth century were the research laboratories of the giant industrial concerns, today the emphasis is increasingly on close collaboration between industry and university research. New software-intensive and knowledge-based technologies are emerging as a result of the progress in molecular biology and in the capacity of computers, which increase university research's direct utility for industry.[30]

In the meantime, it is mostly left up to art and literature to test what ambivalent emotions they can evoke in audiences with, for example, new forms of life created by mutations. Patricia Piccinini's artful figures consciously cross the boundaries between humans and animal species. She creates monsters whose familiar human characteristics disturb and speak to us even while they appear in an alienated form. She arouses the viewers' curiosity, a curiosity that follows two separate paths at the same time. One path leads to the created object, which is alien and familiar to us at the same time. The other path leads back to ourselves, for looking at the object nourishes our suspicion that science's conscious and intentional manipulations could one day also make us humans take on such an alienated form, just as the artist has done with her mutated form produced from silicon, acrylic, human hair, and leather.[31]

Curiosity thus moves between the fronts, popping up first on the one side and then on the other when the pendulum swings and provokes vehement contradiction or resistance. Curiosity prepares for the new without already knowing what form the new will take. Curiosity does not place any limitations on itself; it is deeply amoral (as the church fathers knew, which led them to condemn it). But curiosity does not lead directly to the new and to renewal, and it doesn't like to be steered. Initially, what counts as renewal is what is economically valued and what shows provable economic success. As an equivalent, "objective" nature preserves the market, which is outfitted with "natural" effective powers. Subjectivity is pushed aside to the sphere of the emotional factors that influence purchasing decisions and that trigger feelings of well-being and the satisfaction of needs when a purchase is made. Daedalus and Icarus remain two figures from Greek mythology that ought to remind us that the attempt to advance into the unknown remains risky. The outcome is uncertain.

The Microstructures of Creativity

Curiosity likes interstices where it can move unchecked and that are favorable for the emergence of the new, whose form and point of connection to the old are not previously known. Nor is it clear beforehand what continuities there will be, what kind of ruptures will divide the new from the old, or how the conflict between preservation and overthrow will end. One well-researched kind of interstice is the social site where the individual creative act encounters things and a materialized nature that the gaze can penetrate, the mind can recognize, and the hand wants to change. Since the Renaissance, at the latest, the Western imagination has regarded the creative individual as a creator and the origin of the creation of the new. Hans Blumenberg carefully and precisely traced the long preceding process of transformation and the valorization of curiosity to the striving for and drive for knowledge that was the precondition for the development of the individual.[32]

Researching this interstice, of course, is tricky. After all, the strategy of uncovering roots in the individual psyche or in the social environment in which the creative individual acts is itself a practice that seeks to follow curiosity and decipher it. As clever as the research strategy presents itself as being, ultimately the individual creative act resists being ordered in existing classifications. A look at the numerous biographical and autobiographical testimonies about scientists and artists unearths repeating patterns—for example, how a sudden insight all at once brings together previously separated strands, thoughts, and associations from disparate domains or suggests that they be seen together.[33] The conditions that foster creativity can be listed, including the necessary open spaces and free time. But the final, decisive aspect that could explain the creative act remains hidden in an impenetrable and incommunicable zone. It is as if the aura surrounding the act of individual creativity could be preserved only if it is able to elude its own introspection and with it the words about it. The ladder is pulled up after it is used. If coincidence favors only the mind that is prepared for it, as Pasteur said, then the question remains how this readiness arose in a specific case, and analogies have to suffice as answers to the question of how it can be consciously brought about.

The high value attributed to thoughts, visual associations, and language when following the traces of individual creativity has long concealed that creativity and the emergence of the new are owed not only to ideas and insights but also, to achieve lasting effect, to work on and with things. What the French historian of technology Marc Bloch aptly called "la force créatrice de la chose créé" (the creative power of the created) continues its effects in things. We have already spoken of the cultural matrix, the interacting social group, in connection with the invention and spread of symbolic technologies. For the inventions of the creative individual must be spread and communicated if they are to become part of the social creative process that builds on the creativity that others emit, absorb, and reemit. Language is not the source of ideas; rather, it mediates between thinking and technology. It "washes" (as Marc Bloch put it) what would otherwise be incommunicable and would remain limited to the narrowest local space.

The symbolic technologies made it possible to build up the inexhaustible space of representation for the development of cultural-evolutionary strategies in which replicative information in the cultural memory that is inscribed in things is unloaded outside the limited biological memory. The external symbolic technologies and the networks they form are themselves things in which the creative power of other things and of the ideas that have led to them continue to have their effects. They are the result of a continued history of interaction between things and people who create, mediate, learn, use, and pass on things. Thus, if we ask about the microstructures of creativity, we must always also examine this astonishing mixture, these seemingly inextricable loops and ramifications. By working with delays and temporal jumps, we can at best reconstruct the paths, which are never temporally straightforward, but we can never even partly determine them beforehand. The paths behind us always include paths not taken, as well as stories that would have been possible but that never occurred and were never told in this way. The passage between the space of possibility and the space of reality can be extremely short, and what previously appeared disordered but possible later all too easily appears to have been orderly. This is the famous moment when order arises on the margin of chaos, and this process, irreversible as it is, is also irreversible in the practice of research.

It is no coincidence that recent scientific research and scientific history have turned toward the material and cultural practices as applied, used, manipulated, given significance, and implemented for further practical-scientific work as the "culture in the experiment" in the laboratory[34] but also in the scientific field. Closer examination of these concrete sites of knowledge production has itself proven to be a creative reservoir for observing, analyzing, and commentating on the processes of scientific activity—"emerging science" as well as "emerging technology"—in their temporal and geographical diversity and for conceiving and seeing them from new perspectives. They have placed the materiality of scientific practice—the significance of instruments in producing or reconstructing processes and phenomena that are present or absent in nature—at the center of the deciphering of the creative process. In the laboratory, the context is mostly ignored because the aim is a constant, purified, and siteless environment, but the proper selection of a site with its consciously sought-out particularities is central to field research. In this case, the manipulation of the site seeks to achieve what in the laboratory is striven for by means of the experimental design and by blocking out the site as much as possible. A third site has long since joined these, of course—an imaginary model created at the computer, where the researcher models and simulates what happens in another place.

One of the theoretically most original and historically most precise descriptions of the microstructure of creativity in the laboratory is that of Hans-Jörg Rheinberger.[35] He is interested in those material arrangements that twentieth-century laboratory researchers in the practice and everyday language of biochemistry and molecular biology call *experimental systems.* They determine the framework conditions of research work in the experimental research fields. They are set up so that they can provide currently unknown answers to questions that the experimenter is not even able to clearly formulate yet.

Here and elsewhere, Rheinberger takes up François Jacob's concept of the "machinery for producing the future." In finding and producing the new, the process between the not-yet and the no-longer (which cannot be given precise temporal limits) always points beyond itself. Experimental systems serve to materialize questions and bring forth concepts that embody them. A closer inspection shows that two different but inseparable structures interlock in an experimental system. The first are the objects of knowledge or, as Rheinberger calls them, the epistemic things that are the aim of knowledge's effort. They can be objects but also functions, reactions, or chemical-biological structures. They are still unclear and undetermined; they are "ideas in flux." This state of still being undetermined, of course, should not be seen as a deficit but rather as determining activity. Epistemic things paradoxically embody what one does not yet know; despite experimental presence, they are (still) absent.

To launch the process of operational redefinition that aims to transform the epistemic things from a still vague shape into a defined comprehensibility with clearer contours, either experimental conditions or technological things are needed. These include instruments, recording programs, and in particular the standardized model organisms that are so prized in the life sciences. These technological things, apparatuses, and instruments and the model organisms prepared for this purpose are themselves the result of the technological civilization we live in. "It is not the sciences that have brought forth modern technology," writes Rheinberger. "It was the technological form of life that gave the special epistemic activity that we call science its historical impetus and its irresistible momentum. In the final analysis, the systems of science draw their meaning, dignity, and esteem from this overarching realm."[36] The technological things are themselves sedimentation products of local or disciplinary work traditions with specific measuring equipment, preference for specific materials, or craftsmanlike skills. In contrast to the epistemic things, the technological conditions have to have characteristic qualities in the context of the current standards or purity or precision. They determine the epistemic things in two ways: they limit and bound them, but they also give them space by forming their environment and by allowing them to emerge.

Rheinberger reconstructs this temporally and spatially nontrivial interplay between epistemic things and the technological conditions that allow them to emerge on the basis of a case study devoted to the test-tube system as an experimental system for researching protein biosynthesis. But if an object can function as a technological as well as an epistemic thing, depending on the place that it occupies in the experimental context, and if the interplay between the various components is closely interwoven, then the question of why the distinction between the two things should be introduced in the first place arises if this distinction is subject to a constant historical revision. Rheinberger's answer is unambiguous and brief: the distinction helps us to understand the game of bringing forth the new, the emergence of unpredictable events, and the generation of surprises—for which experimental systems exist. It helps us understand the nature of research.

But the perspective can be reversed if we want to understand the creative power of things. The distinction between science and technology, between the objects of knowledge and the things that help bring it forth, is not identical to the boundary between uncertainty (which is feeling one's way toward new knowledge) and certainty (which is needed to trust that things will function). Nor does the boundary run straight between producing the future and securing the present. Science does not always stand for surprise in finding an as yet unknown identity, nor does technology always stand for its consolidation. Research is not alone in asking questions; technological and machines are built not solely to give answers but also to make additional questions possible. Long before there was laboratory science, for centuries craftsmen and technicians worked as tireless tinkerers. Both science and technology produces the future, and, vice versa, it is also expected that science will secure the present by providing reliable answers from secured knowledge. Epistemic things and the technological conditions under which they crystallize to take shape and achieve a—temporary—identity together form the intertwining alternate strands of the microstructures of creativity. Both are indispensable for bringing forth the new—and yet they are not sufficient to produce innovations.

The Institutional Conditions of the New

Curiosity (to take up the situation in the laboratory), as indispensable as it is, is not sufficient by itself to bring forth the new. The play of possibilities cannot be limited to what happens in the protected space of the laboratory. And between the laboratory and the world outside—a world consisting of the broadest possible spectrum of funding enterprises and firms, stock exchanges, media, defense ministries, and mediation agencies—there are cracks and fissures through which the new wells up and seeks its paths in many directions. The individual researcher working in isolation on her experimental system has been supplemented and replaced by worldwide-networked, collaborating, and competing research groups that, despite constitutionally guaranteed freedom of research, still act within organizations that have to accept institutional directives. Research projects have to be planned and submitted to the appropriate body. A complex system of institutions to promote research evaluates and judges, allocates funds, appraises beforehand and reviews afterward, sets new strategic goals, and thus creates some of the institutional framework conditions that tie basic research to technological innovation and, more generally, that try to target the production of knowledge to make it the motor of economic competitiveness in a globalizing world. The unreleased tensions between scientific curiosity and the institutional attempts to take it and guide it in preset directions determine the dynamics of the emergence of the new.

Objects of knowledge and epistemic things on the micro level of the laboratory correspond on the institutional level to what is called *basic research,* whereby the diversity of expressions like *curiosity-driven, blue sky, uncommitted, oriented basic research* or *fundamental research, science-driven,* and *frontier research* mirrors the differentiated discussions that, since *Science: The Endless Frontier,* Vannevar Bush's pioneering call at the end of World War II, have dominated American research policy in particular. The experimental system, the technological things, find their correspondence in the institutions that are needed to accompany the epistemic things on their usually long and never straight path ultimately to transform them into what, abbreviated as (primarily technological) innovation, is the dominant theme and goal of politics, industry, and targets for economic growth.

The paths open to curiosity are many or even too many, they are never straight or predictable, and it takes different and unpredictable amounts of time to traverse them. This is as true of Rheinberger's case study, which focused on the test-tube system as an experimental system to research protein biosynthesis, as it is of another study on physical and biological developments that provided the prerequisites for the decisive breakthroughs in the diagnosis, prevention, and healing of cardiovascular and pulmonary illnesses.[37] The new biomedical technologies of the 1970s were developed in the course of the increased attention paid to applied biomedicine under the administrations of U.S. presidents Lyndon Johnson and Richard Nixon. Julius H. Comroe Jr. and Robert D. Dripps wanted to know what made them scientifically possible in the first place. The results of this study illustrate in an interesting way how multifarious, differently timed, and nonlinear are the paths of development that emerge from scientific discoveries—that is, from the new that appears in the laboratory and the iterative interplay between epistemic and technological things and that leads to new technologies for treating patients. The example of heart surgery is especially illuminating.

The availability of general anesthesia in 1846 led to an upswing in surgery, except for thorax surgery. Heart surgery did not follow until a hundred years later, and the first successful open-heart operation with a complete heart-lung bypass apparatus was carried out 108 years after the first use of ether anesthesia. What held heart surgery back? What knowledge was lacking before a surgeon could proceed to remedy a heart defect predictably and successfully? Initially, a reliable preoperative diagnosis was needed for each patient whose heart needed repair. For this, selective angiocardiography was necessary. This in turn presupposed the earlier development of heart catheterization, which is based on the even earlier discovery of x-rays. But the surgeon also required an artificial heart-lung machine and a pump oxygenerator that could assume the function of the patient's heart and lungs during cardiac arrest. This machine needed a design that did not harm the blood. For the oxygenerators to function, in turn, fundamental knowledge about the exchange of oxygen and carbon dioxide in the blood was needed. But the most perfect heart-lung machine would be useless if the blood coagulated while it was in use. So heart surgery had to wait for the discovery and purification of a potent, nontoxic anticoagulant—heparin.[38]

Perhaps it was such detailed and impressive case studies that led Donald Stokes to divide the play of possibilities between basic research and technological innovation into four quadrants. Perhaps he took inspiration from the exemplary behavior of a historical figure who mastered the play of possibilities between basic research and application, between epistemic and technological things, and between the individual researcher's hard work and his entrepreneurial talent in managing a laboratory, developing relations to industry, and having a highly developed sense of what is now called PR or public relations.[39] In Pasteur's quadrant, curiosity and the wish to understand are expressed. The search for new scientific discoveries is coupled with the equally strong will to ask about later uses and thus about possible applications.

For a long period, basic research, which is associated with the field of "pure" curiosity and an understanding of the freedom of science nourished by the philosophy of science, was separated from application not only conceptually but also institutionally. The "linear model" on which Vannevar Bush's thoughts are based thus conceives a temporal sequence moving from pure basic research through a phase of applied research to, finally, the commercialization and marketability of developed products. But already in the 1950s, doubts were arising that saw this model less as a valid generalization of the dynamics of the research process and more as the codification of an exceptional episode—namely, the model underlying research efforts in the United States during World War II. Other terms and classifications were suggested that were meant to lead beyond the doubts about whether a distinction could actually be made in practice between "pure" basic research and applied research and whether it made sense for a research policy to promote practical goals and encounter new fields like biomedicine and biotechnology.

The historian of science Gerald Holten took as a model Thomas Jefferson, who proposed and encouraged the extremely successful expedition of Lewis and Clark. Holton considered it an example of "Jefferson's research policy," according to which a specific research project takes place in a field in which there is a lack of scientific knowledge about the foundations of a social problem.[40] President Jefferson, a friend of the sciences whose party had won the elections of 1800, realized that the expedition would have two results. It would be useful to basic research because Lewis and Clark would map the territory and return with unknown flora and fauna and observations about the behavior of the indigenous people of the American Northwest. But Jefferson was at least as clear about the political-practical side of the venture. The westward expansion of the young nation ensured that America would escape the supposedly inevitable fate of Europe, the trap of overpopulation and food shortages.[41] Jefferson was smart enough to emphasize the commercial side when touting his project to the Congress as funding body. To the Spanish authorities, which controlled part of the territory that the expedition passed through, he underscored the scientific aspect. Indeed, the mixture of both motives was decisive: he wanted to promote the best research, which promised no short-term use at all but which would take place in an area central to a recognized societal problem.

In Pasteur's quadrant, which has an undeniable similarity to the geometry of the Newton-Bacon-Jefferson triangle used by Holton, two dimensions also come together. One is borne by the search for understanding of the foundations and is thus on the trail of the epistemic things, while the other has its eye on possible applications or their further use. Two other quadrants are devoted to Bohr (for whom practical use is irrelevant) and Edison (who is interested solely in practical use). The fourth, unnamed quadrant consists of systematic work on specific, particular problems that do not aim at either general understanding or application but that are part of a greater whole.

Today, the public's interest in research policy is hardly at the center of attention. This was not always the case, even when research policy had more the character of an expedition or was a program auxiliary to military campaigns. The contemporaries of Jefferson and Pasteur, Karl Linné and Joseph Banks, Alexander von Humboldt and Charles Darwin, understood very well what these expeditions served. The greater program consisted in the expansion of European colonial empires or served the Western striving for expansion in other ways. The contemporaries shared the fascination this created and probably also the economic and political goals. An expedition that set off into distant lands would retrieve nature and everything there was out there and settle it in gardens at home or bring it into the laboratory, where it would be dissected and magnified, taken apart and put back together again. The knowledge of how things functioned then served to bring newly produced or changed technological things, as well as plants and organisms, chemically synthesized substances and medications, from the laboratory to agriculture, gardens, hospitals, markets, and also future battlefields.

Pasteur's quadrant and Jefferson's research policy are more than classificatory exercises. They influence the direction in which scientific curiosity is guided and the ways that it can serve societal forces that are today soberly described as politics, economics, and society. They make clear the degree to which scientific, technological, and institutional practices are inseparably mixed with the local and temporal coordination of the contexts of their application. Scientific curiosity and the knowledge it produces are always situated, and yet they change their form, even if nature's laws or certain principles of its functioning remain the same. Concretely, knowledge adapts to the production site, which can be a university or industrial laboratory, a start-up company, or a consulting firm. The dynamics of knowledge production and of the growth of knowledge, the paths that scientific-technological curiosity must traverse to transform itself into innovations, are always multiple and never straightforward. The course they take runs from the first public announcements of their existence in scientific journals or in a patent through various material apparatuses and infrastructures to spaces in which the knowledge is tested and elaborated and collaborates or competes with other, locally created knowledge and technological things. It circulates through channels of exchange and information, diffuses into the many heterogeneous sites and contexts of use in which it materializes and finds temporary stability, until the next large or small wave comes, bringing change and expansion with it again.

The internal logic of scientific-technological curiosity sometimes begins with newly formulated theoretical or experimental questions or continued work on questions that others have left unsolved. But "pure" motives are rare because other impure purposes and goals already join this first logic. They can be experimental or instrumental—that is, still limited to the world and the work of the laboratory, the workbench, and their experimental systems. These in turn are inseparably tied to the logic of the institution that knows its aims, even if they may be only coarsely defined: just as Lewis and Clark knew what was expected of them when they sketched maps of the territories they surveyed and when they put together as well as possible an overall image of the country with its human, animal, and vegetable inhabitants, which aimed to express a claim to control.

Pasteur wanted to understand as well as control the microbiological processes he had discovered. And to mention the social sciences, John Maynard Keynes wanted to understand how a modern economy functions on macro- and micro-levels and at the same time to develop a toolkit that would be suitable for interventions. For the generation of physicists and engineers that, in the framework of the Manhattan Project, worked feverishly to build the atomic bomb before Nazi Germany could do so, the first step was to learn to understand the newly discovered processes of physics. Nonetheless, they were driven by the desire to implement this knowledge in practice immediately. Modern electronics requires comprehension of the processes of surface physics, just as molecular biology builds on the growing knowledge of genetics and proteomics to be able to intervene in, manipulate, and change natural processes.[42] Under the respective institutional conditions of production, insatiable curiosity is not permitted to limit itself to understanding. Curiosity leads to changes; it includes them even if they are not completely known. Scientific-technological curiosity thereby creates new epistemic and technological things, new questions and instruments. To persuade others and to receive the necessary material resources, it creates not only instruments but also institutions whose logic it knows how to use for its own purposes. It creates contexts of application and spaces whose institutional practices combine with the practices of the experiment or the calculation in that logic of dual use that brings the world into the laboratory—to change the world from the laboratory.

The Wish to Control the Unforeseeable

Technology also knows many paths—and many paths not taken—that lead from the first inventive idea through countless hurdles and selection criteria to the successful product or process.

Depending on the site, the search for ideas on the micro level can be distinguished from the implementation and development of large sociotechnological systems. Inventors' search strategies have been compared to "Klondike spaces"[43] (a metaphor borrowed from the Yukon gold rush) in which the coveted gold is where it is found—sparsely distributed and without clear indications of where one should look for it. But Klondike spaces are typical of problem-solving and search strategies in situations that demand sudden insight, as in human inventions and biological evolution. The suddenness of recognizing and finding, the breakthroughs in understanding or producing, and the punctuated equilibrium of evolution in which sudden transitions in adaptive form occur are characteristic examples of similarities between human discovery and invention, on the one hand, and the strategies underlying evolution, on the other. Human constructors can work with ideas and concepts that lead to alternative sketches, which in turn can be developed into competing prototypes whose emerging problems lead to revisions not only of the sketches but also of the original concept. In addition, a poor concept can lead to a good one, and a prototype that does not function can lead to a functioning one. The limits of the possible, however, are more pronounced for biological evolution than for human inventions, which are able to vault over them in a leap of imagination. But nature has at least two advantages: its time scale is much longer than that of human existence, and it can carry out massive parallel operations. Up to now, nothing invented by humans could keep up with this.[44]

In the laboratory, the new finds entry through cracks and fissures and leaves again in changed form if the epistemic and technological things are to have further effect. Similarly, the technological solutions and prototypes, the models and sketches, the designs and architectures developed in computer models ultimately must find a way outward to find their place in the technological macrosystem[45] or in a niche within one or between several such systems. Characteristic of large-scale technological systems is that their technological dimension is integrated with other economic, cultural, or political dimensions. Traffic systems, for example, require not only the necessary technology but also investments, infrastructures, and regulatory systems. They are based on cultural presuppositions, like the value attributed to collective or individual mobility and preferred energy sources. Elaborate delivery systems and infrastructural facilities must be present. Well-planned control and information systems are needed to secure coordination. Historians of technology[46] describe the emergence and development of such large-scale technical systems as the course of "technological trajectories." These generally leave several possibilities of further development open at the beginning, whereby chance constellations can often decide what direction is then taken. There are repeatedly bifurcations. But to the degree that one of the technological trajectories prevails, it gains stability and entrenches itself. It can grow to the point of impermeability and irreversibility, thus achieving a state described as a rigid, "locked-in" technology. Every change is accompanied by high costs. Technological innovations, if they are to prevail, must thus either find their place in one of the macrosystems or themselves belong to an innovative technological cluster that will grow together to produce a macrosystem.

Nothing in the precise historical analysis of the various paths that curiosity and innovations take within and between institutions indicates a technological determinism, nor are there lines of scientific development that follow one and only one logic. The motto for such developments could be, "This is how it is; it could be otherwise."[47] But precisely this openness is the source of the problem of society's dealings with the new. If the paths that curiosity will iteratively take or overleap in the laboratory between the objects of knowledge and the technological apparatuses can be neither planned nor foreseen, and if the Klondike space lures gold seekers who are uncertain about how much and where gold will be found, then a paradoxical double game ensues. On the one hand, focused efforts will be made, and new plans drawn up. Strategic goals will be put in new order or defined by foresight exercises, whose goal is to guide the new, which is either not yet present or is known only in vague contours, into certain directions. On the other hand, it even appears that the more the new is able to elude intention, the more planning and strategies are employed to capture it.

This results in a dilemma for every institution that seeks to foster the new: it wants to bring forth the unforeseeable and yet keep it under control from the beginning. It seeks to find applications and uses that no one knows yet, and at the same time it seeks to eliminate or minimize unintended side effects and possible risks. The new, which initially always arises only locally, should spread by the path that leads through new contexts of application to the market and to societal and economic uses. The double game consists in optimizing the production of the new and then being able to put it all the more selectively into specific forms and send it in specific directions. There are as many selection filters as there are economic and legal framework conditions, organized resources and interests, cultural values, politically organized spaces, and various forms of organization. Added to this is what already exists. It has a power of persistence based not only in values but also in the things themselves. Successful technological innovations therefore often make use of an already existing form, thus ensuring they can function compatibly.

The technocratic certainty claimed to know society's needs, which was systematically organized to fulfill state-supported research. It was manifested in many large-scale projects, whose realization was in no way restricted to totalitarian regimes. The confidence of a worldview that thought it "could see like a state"[48] laid its comprehensive claim to shape the future that it asserted it could foresee. The irony of history is that many of the utopian visions of the future from the extremely politically and ideologically charged period of the 1930s have been realized or outdone today. And yet there is no perceptible slacking of efforts to use research to drive forward economic growth. The thread running through it all, which the British crystallographer and Marxist John Desmond Bernal presented in his 1939 study of the societal function of science, [49] is timelier today than ever before, though under the opposite sign of a neoliberal ideology. There is broad agreement that more money should be invested in research (that is, that science and technology must continue to expand) so that society's affluence and well-being can increase. This is to be achieved by putting the unexpected and new that comes out of the laboratory into the widest possible variety of contexts of applications to produce in them new knowledge that in turn brings forth new abilities and continues to spread in society. The statist viewpoint of "seeing like a state" was replaced by the neoliberal credo that attributes newly discovered, sensual qualities to the market. Now it is the market that claims to know people's most intimate needs and the relationships between technological things and between people and things and that has discovered in the omnipresent network structures the most efficient structure of organization.

To take up Rheinberger, Jefferson, Pasteur's quadrant, and Bernal (who thought science requires curiosity but that curiosity did not lead to science): curiosity is indeed not enough to bring forth the new or to win the collective bet that all highly technologized industrial societies have made on innovation. Behind it lie the void of an uncertain future that is to be filled and the competitive pressure exerted by many actors costaging a worldwide globalization (which in Greek bears the lovely name *pankosmoiopoise*). Curiosity nourishes itself on questions that point beyond or cast doubt on the usual and given to explore what lies beyond the obvious. It thereby resists premature commitment. It clings as long as possible to the playful and uncommitted impulse, whereas innovation already has allegiance to the introduction of the new and its integration in the given, even if this means the abandonment and disappearance of what already exists. As we have seen, scientific curiosity requires the interlocking interplay between epistemic things and technological conditions, as is present in the laboratory. But the attempts to tame this scientific curiosity, which begin soon thereafter, demand an extended perspective that leads outside the laboratory. The experimental systems do not remain isolated in their laboratories, nor can the laboratories be isolated from the surrounding societal, political, and economic contexts. What happens in the laboratories sooner or later feeds the innovation systems that have formed around them institutionally.

In the middle of the nineteenth century, when university and nonuniversity research laboratories emerged and the specialization of disciplines took shape at the universities (which distinguished for the first time between pure and applied research), the systematic search for the discovery of the new and for inventions was institutionalized. The space where curiosity could unfold achieved its critical density and a stable institutional framework. It was also granted the freedom to move in directions that cannot be pinned down in advance. Today, these institutional framework conditions are changing again at a speed that is redefining the existing rules, directives, and limitations. In this way, the public character of science is changing.[50] The play of possibilities that was once curiosity's very own game is becoming the game of many players, a game of innovation. Curiosity is challenged to continue its game under changed conditions that, on the one hand, want to give it directives and that, on the other, encourage it to continue acting subversively.

Once objects of knowledge and technological artifacts have initially formed in the narrow space of the laboratory, the computer screen, or the local workshop, their initially fragile reality stabilizes by means of their ensuing mobility and the new configurations into which they enter. The knowledge brought forth by the new becomes knowledge of the new. The epistemic-technological mobility of things then takes them to expanded spaces. These can be the spaces of societal and technological networks, but they can also conjure up that world of flows of objects, people, ideas, commodities, images, information, and ideologies that are all kept in motion by the processes of globalized unbounding.[51] But even this flood of movements at some point leads to spaces of activity and negotiation—populated by a large number of actors with different interests, cultural resources, and values and with imbalanced power.

At this stage of realization and implementation in the various layers of societal and political reality, additional demands are made on the new's stability, robustness, and suitability to be integrated in existing structures, technical objects, and their use. The initial fragility of the objects of knowledge that have not yet found a materiality that is emerging or intended for them makes way for a fragility and precariousness of another kind when they enter layers of reality other than those of the laboratory. The new is now to be integrated in society and made compatible with what exists. This entails conflicts, and resistance forms. Adjustments must be made; complex, already existing organizational arrangements must be overturned and rearranged. The scientific-technological components of the innovations can no longer be cleanly separated from the economic and cultural ones. Suddenly, the previously concealed nonsimultaneity of the old becomes visible in its historical constructedness. The old loses its privileged status of self-evidence and must defend its vested interests. This leads to a growth in the pressure to manage this nonsimultaneity with the old by taking recourse to the simultaneity of the new, which makes its appearance in the imperative of the present. New uncertainties thereby appear and demand decisions that earlier were irrelevant or not desired. The epistemic fragility of the new demands a social robustness, new social forms, and its own language to persist in the societal space.[52] The more that society depends on the technological-scientific culture it produces, the more robust and at the same time more vulnerable society itself becomes.[53]

Orders of Knowledge Are Social Orders

At the beginning of the twentieth century, Emile Durkheim, one of the founding fathers of sociology, investigated the connection between religion and society. Religious phenomena presuppose a division of the world between the familiar and the unfamiliar, between the holy and the profane. What is holy is protected by prohibitions and isolated from the profane. Profane things are those that are affected by these prohibitions and that must be kept at a distance from holy things. Systems of religious belief are representations expressing the nature of the holy and relationships to the profane.[54]

With its historical rise, modern natural science took on some of the social functions that were formerly the provenance of religion. The distinction that science makes between the person who knows and the person who is ignorant is analogous to religion's distinction between the holy and the profane (the

French word for laymen is *les profanes*). The sacred centers of science are to be found in the precinct of science, which is considered untouched and beyond the influence of any human action or intervention. This is the hard epistemic core of natural science[55] that is grounded in the laws of nature and provides the basis for the conviction that this core knowledge is about nature's laws. This knowledge is the source of the collective force of the natural sciences and contains the social energy to unite its members. It is the basis for epistemic and social authority with which the scientific community speaks in the name of a higher order of knowledge that is beyond human control.

From this firmly anchored conviction grows natural science's characteristic and sometimes vehement resistance to all attempts to uncover the social roots of its order of knowledge. Their existence, of course, cannot be denied because it is incontrovertible that science consists of a great number of societal and cultural practices. It is equally uncontested that science is a social institution that stands in a complex mutual dependency with its respective society. The production of scientific knowledge is thus always part of a comprehensive process of contextualization.[56] But what is contestable is how great a role the social order plays in the order of scientific knowledge and the way that the latter depends on society in the production of new knowledge. Like other creative activities, science needs free space for curiosity and curiosity's pleasure in experimenting. Science claims to have found a privileged access to the order of knowledge that is removed from the grasp of every social order. So to protect its access to this higher order of knowledge, it must minimize its part in the social order. It must insist on the strict separation between a holy order and a profane order, thus preventing an unwanted and unacceptable contamination by the social order.[57]

Nonetheless, despite all productive misunderstandings and sometimes destructive controversies and despite all understanding for science's defenses against permitting the normative, societal order to have more say in science's activities, science and the social order are interconnected. Every society is also an order of knowledge and is based on the existing knowledge that its members have from and about each other and about their common natural and social world. This interconnection does not mean that the one order can be reduced to or is the cause of the other. In particular, the production of scientific knowledge cannot be reduced to what is "socially constructed," even if investigations of the subtle influences of societal framework conditions, questions, and the framing of the questions and of research priorities have shown that the processes of producing knowledge are not independent of their societal context.

There is in fact something that is present "out there" in a reality we perceive as objective, that does not answer to our desires, and that is only partially accessible to our interventions. But that we know this and how we know it would be impossible without the social order. Human curiosity is itself the product of nature and society—a mixture of biological, neurological, and cognitive preconditions, processes, and cultural practices that have arisen in the course of biological and cultural evolution. The insatiability of curiosity made it possible to bring forth the new and to institutionalize this dynamic culturally, economically, and politically. The speed of today's developments calls for a culture of dealing with the new and its innovations. It demands that we once again think of the order of knowledge and the social order together in their mutual dependency and interlocking.

One of the diffulties of describing the transitions—from the level on which untamed curiosity begins acting, through the emergence of the new (which begins stabilizing in the laboratory), to the level of the macrosystems (whether large-scale technological systems, the global economic system, or "society") where the new unfolds its effect as innovation—is that what is initially invisible must be made visible. The process of description turns the unexpected and unforeseeable into something that now seems possible to foresee and plan. In the process of description, which reconstructs its genesis, what cannot be (because no one has ever conceived or imagined it) becomes something self-evident that everyone always already knew.

Description serves to make explicit what is initially implicit.[58] Description names the new, gives it a language (without which it cannot become socially mobile), and provides it with understandable attributes. In short, description gives the new a social form. But the phenomena that are described are never merely a product of the imagination—even if imagination is always in play in them. They are there, already scientifically graspable though they have just arisen and become. Their emergence is closely interwoven with the process of describing and with their social perception. In this sense, the knowledge order and the social order meet. In the process of the emergence of the new and in the process of describing this emergence, the two orders are interlocked.

What plays out within an experimental system in the laboratory first becomes interesting when there are results. Describing the routine, the frustrations, and the failures interests us only as part of a story that leads to a good or bad ending. In the case of the laboratory, the results are objects of knowledge that stabilize enough to be reproducible and that other researchers in other laboratories can observe, produce, describe, measure, and make serviceable for further experiments. Making what happens in the Klondike space visible becomes interesting and worthwhile only if one of the search strategies used shows success and if the gold seeker finds something between the "plateaus," "canyons," and "oases," puts her insight down on paper, or presents it in another way that makes it understandable to others. Only then can evidence be brought that the idea, the concept, the design, or the prototype was suitable to solve one of the problems posed. The process of searching for gold, the inspiring idea, or the brilliant solution is interesting only because it has led to a result that others can take part in. Making the result visible is indispensable so that the new can be taken up, used, appropriated, and changed by others. It is the precondition for enabling the social order to be creatively active.

The process of describing and naming must mediate the temporal discrepancy between what is not yet, what is becoming, what is in the process of becoming, and what is present and perceived as new. It is itself process and must reproduce a process that played out independently of it. It must do this with means and in a register appropriate to itself, just as the process of the emergence of the new has its own proper register. What is becoming is still invisible or skillfully hides behind the old. It quietly does its work, and even if it is sometimes announced loudly and shrilly, it never announces itself loudly and shrilly. The emergence of the new is inseparably tied to people because new ideas and insights arise in and through individuals. But the meshwork of their relationships to others, the conversations connecting them to others, and the coincidences that take them to certain places at certain times are equally still invisible and inaccessible to description. Their reconstruction is difficult and can never be complete, even retrospectively. It is no coincidence that we do not encounter any people in the experimental systems of the laboratory. The laboratory as a local site and social organization does not interest us until the result makes its public appearance. Then the laboratory receives its place in a scientific publication, where the names of the authors and the site where they carry out their scientific work are listed. The social order is needed to have the order of knowledge appear.

Similarly, the social order becomes visible only gradually in the process by which the new emerges from the order of knowledge. It takes on significance only to the degree that it provides results that are socially and culturally esteemed, taken up, and further used. Fully developed, consolidated, and exerting their powers, they appear in the macrosystems. In them, the selection made can be perused. The respectively dominant selection criteria have done their work. The economic and political driving forces that are at work show their effect. Preferences are articulated, set directions, provide incentives, and sanction what is not desired. Constellations and conflicts of interests must be regulated, and transparency and a rendering of accounts are increasingly demanded. The new knowledge and the new technologies then appear as shaped by society and adjusted to an economic logic or other purposes and goals. Curiosity at first acted within a more or less responsibility-free space, but every macrosystem seeks to gain and preserve control over its innovations. The unforeseeable that it tried to lure and stimulate is now mercilessly checked for side and long-term effects. The social order tries to deal with its inherent vulnerability by carrying out risk analyses or by relying on the precautionary principle. The shaping power of what has now become visible is overestimated, just as the previously invisible was underestimated. The description of the new tends to be socially overdetermined.

The connection between the order of knowledge and the social order may become clearer if we conclude by changing yardsticks, spatially as well as temporally. The laboratory, in which epistemic and technological things are interlocked, tells a story of traces and things. It should be supplemented with a story of cultures and things. Of course, the historical and empirical material thereby remains fragmentary, and the categories that are used remain coarse and inadequate. So an urgent warning against premature and inadmissible conclusions is in order.

The Cultural Diversity of Curiosity: The Word and the Way

In the period between the fifth century BCE and the third century CE, impressive scientific and technological achievements were made in the two great civilizations of the world at that time—in Greece and in China. In their joint work, *The Way and the Word*, the historian of the ancient world Geoffrey

Lloyd and the sinologist Nathan Sivin ask about the specific configurations in which the curiosity to know and the ambition to create something new were anchored in social institutions and about the results they led to.[59] Of course, here we cannot speak of science in the modern sense, and the development of technology took its own path, independent of science, as it did everywhere until the nineteenth century.

Comparing Western and Chinese civilizations has itself a long history that has been shaped by cultural mobility, the fascination of the exotic and the other, as well as numerous productive misunderstandings. One of the leading questions in the history of science and technology that already greatly concerned Max Weber and Joseph Needham was why the so-called scientific revolution occurred only in the West and thus remained a singular historical process. This question must remain unanswered. The second leading question explores the social, political, economic, and cultural differences that explain why China had a leading role in many technological areas but the West nonetheless caught up with and overtook it. This question can be answered only with much differentiation and on the basis of many individual studies.[60] Lloyd and Sivin have chosen another way. They explicate commonalities and differences and then examine the relationship that the creative individual had to his social group and the influence that the social organization had on creative achievements. In this way, they reveal the roles and functions of institutions in the production of the new in the two societies.

Curiosity thereby appears in a kind of primal shape—socially still mostly formless and naïve but equipped with the power that is already inherent in the genesis of symbolic technologies. In earlier civilizations, curiosity seemed to be driven by a wish towering over all others—the desire to be able to make predictions about the future. Numbers were regarded as the key to understanding phenomena and systems because, if worked out in adequate detail, they made predictions possible. In ancient Greece, people believed they could see numbers in things, which led them to analyze physical phenomena with the aid of mathematical models. The ancient Chinese, by contrast, displayed no ambition to derive the whole of mathematics from a few axioms. Their goal was to use mathematics to better understand the social order and especially the unity in it they strove for. In China as in Greece, numbers were used to illustrate societal systems of order, and in both civilizations, people who were able to present themselves as experts in the manipulation of numbers enjoyed high social prestige. The Greeks, however, regarded the mathematically conceived universe as, in principle, independent of human beings. It was considered objective in the sense that this order could not be contradicted. The Chinese saw mathematics as a source of social cohesion. It functioned as a symbol of the unity of the empire, which was striven for and repeatedly put into question.

There are also interesting differences in the practical application of knowledge and in the amount of theoretical knowledge that was drawn on to this end, although the use of the terms *useful*, *practical*, and *theoretical* must remain problematical. As is well known, stereotypes are long-lived. The stereotype of China's relative technological superiority finds its correspondence in the stereotype of contempt for manual labor in ancient

Greece. Although there may be a grain of truth in this, there is an astonishing wealth of counterexamples that call for a more differentiated view. Lloyd and Sivin repeatedly underscore that, especially in the history of technology, the state of the primary sources has many lacunae and that generalities often merely display an author's preferences. The remaining differences are based on the two civilizations' differing cosmologies. In Greece a predilection for geometrical idealizations prevailed, whereas in China the focus of investigation was on the properties of things. The goal was not to master matter but to search for ways that enable people to cooperate with nature and win it as an ally for their purposes. A similar approach is known, incidentally, from Chinese military strategy—for example, when (as in the game of Go) the aim is not to annihilate the opponent but to induce him to surrender. The schema for acting is intensely situational. The point is to grasp possibilities that are given in a concrete situation.[61]

Such glimpses of the differences in how people deal with nature, use schemas of means and ends, or place importance on planning in contrast to a greater readiness for situational action may tempt us to attribute them to an unchanging "worldview," but it is difficult to adduce stringent empirical evidence for this.*

* Richard E. Nisbett, *The Geography of Thought: How Asians and Westerners Think Differently . . . and Why* (New York: Free Press, 2003). Lothar Ledderose remarks: "There seems to be a well-established Western tradition of curiosity, to put the finger on those points where mutations and changes occur. The intention seems to be to learn how to abbreviate the process of creation and to accelerate it. In the arts, this ambition can result in a habitual demand for novelty from every artist and every

Doubtless, a society's own cultural cosmology shapes the images that it makes of knowledge as well as its concrete work on knowledge about and in things and practical procedures. Cosmologies should be understood as systems from which one cannot break out a piece at will to treat it as essential and unchanging. Cosmologies by themselves do not create transport systems, nor do they contain instructions for processing salt. They also have little to contribute to explaining why in Greece the art of rhetoric enjoyed great prestige and inspired the wildest theoretical speculations, whereas in China the highest aim in astronomy was to make observations that were as concrete and verifiable as possible.

To approach such differences in culture and things, we must compare the social orders and their structures. Institutions can foster or inhibit the creativity of individuals as well as of groups. They create incentives that can have positive effects, or they erect barriers that immobilize ideas and persons and lead to stagnation. Institutions are the mediators that either meet the promiscuity of curiosity halfway—or hinder it. To oversimplify: the bureaucracy of ancient China's central political power watched over, guided, and made use of all creative and

---

work. Creativity is narrowed down to innovation. Chinese artists, on the other hand, never lose sight of the fact that producing works in large numbers exemplifies creativity, too. They trust that, as in nature, there always will be some among the ten thousand things from which change springs." Lothar Ledderose, *Ten Thousand Things: Module and Mass Production in Chinese Art*, The A. W. Mellon Lectures in the Fine Arts, Bollingen Series XXXV (Washington, DC: National Gallery of Art, 1998), p. 46.

innovative activities. In China, organizations formed very early that, as precursors of ministries, were assigned to carry out public works, agriculture, jurisprudence, and warfare. Acceptance in the civil service offered desirable career opportunities at the royal courts, temples, or other state services to a broad spectrum of social groups. Employment tied so closely to the power of the state meant that the imperial court and its ministers were the primary addressees and clients of the knowledge that was produced. This doubtless led to a certain dependence on political power holders and increased pressure on the producers of knowledge to respect the authority of the canon of knowledge they jointly created. The painstakingly cultivated external image of unity permitted internal dissent but only within clear bounds. In addition to civil servants, knowledge producers could also work in freelance professions as physicians, architects, engineers, astrologists, and teachers—of course, on condition that the services they offered found purchasers who were prepared to pay for them.

In contrast to this, there were very few established positions in ancient Greece for those who produced new knowledge and who had to offer their technical-practical abilities. It was a matter of individual skill whether a teacher was able to prove his mettle in public speech. Competition was intense, both within a man's own group or school and between the schools. Public reputation grew with success, which was won with public debates and skillful argumentation and the performance of the high art of rhetoric. For Lloyd, the ancient tradition of publicly conducted argumentation in debate is one of the key institutions enabling us to understand how knowledge could develop in Greece—quite comparably to the institution by which the Chinese imperial court and its bureaucracy fostered the acquisition and dissemination of knowledge.

The institutional framework conditions that reigned in the area we are familiar with in Greece—philosophy—were not the same in other areas of knowledge. Astronomy had different presuppositions from medicine, while yet others were considered valid in agriculture or cosmology. Both societies experienced repeated periods of swift, dynamic development that alternated with periods of relatively long stagnation. But state support for and the resulting greater dependence of astronomers in China, who as civil servants also had to carry out religious-political functions, did not prevent them from acquiring remarkable knowledge. In contrast, Greek astronomers received no institutionalized support at all. They enjoyed complete freedom to formulate their own questions and to shape their research as they pleased. But the lack of a corresponding institution often meant that their results would be simply ignored, isolated, or forgotten. No team of state servants waited for them to carry out astronomical observations, as was a matter of course for the Chinese astronomers. Perhaps it was precisely this lack that led them to indulge again and again in frequently interesting but wild speculations.

Lloyd and Sivin draw cautious conclusions from their comparison of the cultural traces that curiosity and things have brought forth. Both societies had institutional structures that, intentionally and unintentionally, fostered the production of new knowledge and of technology with impressive results in many fields that were of great practical and theoretical relevance. These institutions made it possible to create the new, but the new knowledge and the technological things also influenced the continuity of the institutions. Worldview, ideological ideas, social structures, and concrete procedures in fields of activity that have meanwhile come to be called *science* and *technology* interpenetrated each other in China and in the West. The Chinese yin/yang principle, for example, which is the foundation for understanding the Chinese cosmology, served both to legitimize imperial rule and also to strengthen it. The other way around, the pluralism inherent in the Greek ideals of acquiring knowledge directly influenced Greek cosmology. Speculation was not merely permitted but was desired and stimulated a fruitful production of theory.

Both societies had a wide range of institutional support. Considering the impressive results, however, neither of the two civilizational macrosystems can be judged to be the other's superior. In China, the state's bureaucratic structures offered lifelong employment that made a continuity possible that was not available in Greece. Continuous support was absolutely necessary for major technological projects with practical significance. The other side of the medal was that talented individuals were constantly in danger of losing their livelihoods if they fell out of political favor. This is the source of the pressure that constrained diversity of opinion to avoid jeopardizing the external appearance of unity. In Greece, by contrast, a confrontational, pluralistic, and competition-promoting model of social order flourished. To make a name for oneself, it was not enough to be better than one's rivals; one had to be publicly perceived as better. But the considerable individual achievements often lacked continuity.

Posterity's judgment is always post facto and seldom final.

And yet it is amazing to see the apparent return of basic institutional patterns. "The way" and "the word" offer different answers to how, at the beginnings of civilization, societal institutions helped shape the acquisition of knowledge and the production of the new by individual persons who belonged to a clearly identifiable social group. The comparison offers glimpses of the mutual interactions that appeared among the order of knowledge, technological activity, and the social order. In a world where technological artifacts and infrastructures are regarded as part of the social order, it is easy to overlook the intensity of mutual dependencies and interdependencies. Every social order comprises an order of knowledge, and every distinction between "nature" and "society" is made on the basis of images of knowledge that circulate in the society. There is never the one, single, right way—or the one right word.

3 Innovation in a Fragile Future

What does it mean? What does it mean?
Not what does it mean to them, there, then.
What does it mean to us, here now.
—W. H. Auden, The Orators

The Past as Future

An innovative idea is recognizable by the fact that it surprises. The greater the surprise, the more innovative the idea.[62] But innovations do not consist solely of ideas, even if ideas are where they start from. Innovations are tied to the respective context. They consist in the recognition and implementation of new possibilities that reach beyond the tested or accustomed routine. They are defined by their success, which consists in opening up new spaces for activity, whether in connection with technological products, new markets, organizational adjustments, or other social arrangements.[63] The surprise they can trigger no longer comes from the idea but from the effect they can have on life and work, on accustomed ways of seeing and thinking, on feelings and seemingly deadlocked arrangements and power structures. Surprising is also the speed with which an innovative idea can turn into an innovation and the speed with which an innovation can spread and change an existing situation. Innovations blur the boundaries between the present and the future. In many areas, the dramatic changes have pushed open the door to the present for new demands and possibilities that can be expected from the future. Like a breaking wave, the new communication and information technologies pour into everyday working life, where they destroy jobs or force the outsourcing of skills. The speed that has come with electronic data transfer has also increased the pulse rate of the present. To live *for* tomorrow means already living partly *in* tomorrow. One seeks to make oneself fit for the next techno-ecological niche that humankind has constructed with the aid of its science and technology.

Of course, there is also the countermovement. Politics and social movements do not mobilize so much by means of designs for a utopian future as by recourse to the historical legacy, fundamental values, and religious ideas. The closer the future seems to approach the present, the stronger the past's power of attraction proves to be. *Lieux des mémoires* (sites of memory) are set up, historical identities are invoked, biographies and remembrances boom. Historians report that today their discipline not only explores dimensions of reality that were inaccessible to earlier historians but also can more comprehensively perceive and judge the long-term effects of, for example, the nineteenth century than it could in the past.[64] The gaze backward into the past is extending, not least also because of the scientific-technological instruments available today. Ice core samples in the Antarctic or Greenland permit conclusions about the long-term shift in the climate, and a new generation of satellites in space permits a glimpse all the way back to the beginnings of the universe.

Nonetheless, the increased sensitivity to the changes in historiography and the reconstructions of human life in the past as it may have developed through periods of climate change in prehistoric times[65] still do not permit us to deduce any convincing perspectives on the future. The ideas of the future today have become more fluid, elusive, and volatile. Where the predictions of the natural sciences find support in more or less secured data and models, in many areas we must still reckon with unpredictable human behavior. Whereas past images of the future had a common if also utopian site, today there are hardly any overarching utopian models. The future as a collectively yearned-for space, occupied by wishes and expectations, must fail due to the plethora of what seems scientifically-technologically feasible—never mind what parts of the latter are desirable. Have we lost our future because it already claims us too much in the present?

History shows that there are many different ways of imagining the future. Long before the project of European modernism, there was the opposite movement in history, an incursion of the past into a datable present out of which ultimately grew an undreamt-of, productive future. It happened in a place and by means of a discovery, but neither the one nor the other can explain the resulting effect. In the middle of the fourteenth century, Petrarch set off on a stroll in medieval Rome. In a letter to Giovanni Colonna, a member of the Dominican order, he describes his impressions, depicting the ancient monuments and his own feelings on seeing them. His elucidations are accompanied by literary passages from the texts of ancient authors. Both the description of the monuments and the quotations from the old works are suffused with his thoughts about the heathen and Christian world and its wisdom. For Petrarch, the ruins that the inhabitants of Rome ignore or had long since forgotten suddenly took on a significant new shape. They provided the physical and literary material serving to found ancient Rome anew or, as we would put it today, to reinvent it.[66]

Thus began the project of the Renaissance, which spread in the following two centuries and unfolded its unrivaled creativity. It was based in a return to a past seen with new eyes and perceived anew with other senses. The ruins of the ancient architectural monuments in Rome ceased being ruins. Their description provided the basis for attributing new meaning to them, for bringing them alive and into the present. The old texts provided the knowledge needed to resurrect the ancient city with its monuments and statues. First, new texts were written as commentaries on the old, but soon one commentary was written about the last. The knowledge of ancient knowledge spread. It grew with each new elucidation written about it. To literature was added knowledge about practical arts and the mathematics that contributed to ancient architecture, knowledge of mechanics as well as of poetry and cosmology. The distant past came closer to the present and began to coalesce with it. Humanism received an institutional foundation. It organized itself as an artistic and intellectual movement that allied itself with the ruling strata and brought motion into the rigid social structures. Its program was oriented toward the paradoxical appearance of "a passionate hope directed toward the past" and toward the "vision of a new world built upon ancient texts." The Renaissance, wrote historian François Hartog, had an unprecedented ideal (and an unprecedented site) for its audacity—the entire knowledge of the ancient, pre-Christian world. Its courage consisted in appropriating this past and in daring to make a new beginning in the present.[67]

Today, the entire knowledge of humankind and its impressive technological capacities is oriented toward a future that does not so much promise a new beginning as a further intensification and dynamic continuation of what has already been achieved. Science and technology cross the threshold between the present and the future unhindered, for what appears possible in the laboratory today can already be in the market tomorrow or the day after. The scientific-technological gaze thereby literally goes way back and way ahead. The most recent generation of very large telescopes and of space missions like Galilei and Cassini permit scientific curiosity to reach distant galaxies and regions of the universe where its gaze is directed into the past. Molecular-biological technologies permit research on and comparisons among the evolutionary pasts of organisms; robot probes sent to Mars explore the characteristics of its surface to draw conclusions about its origin and development and to compare them with those of the earth. The exploration of outer space, perhaps humankind's future home, leads via the past back into the future. Instead of investigating ancient architectural monuments in connection with the literary testimonies of a Titus Livius, today inter- or transdisciplinary connections are created between physics, biology, computer science, mathematics, and other areas of knowledge that aim to launch new fields of study with the aid of technological instruments and methods. Networks and social organizations are invented and new and old units are founded so that everything that arises locally can be distributed in a way that, as in the Renaissance, allows knowledge to be carried out into the world, where it enters into new combinations.

The place of *renovatio*—the refounding of an old, admired order, its stock of knowledge, and the newly discovered residues of its former glory—is today taken by the *innovatio* that these modern societies that are willing and able to change have devoted themselves to. The new is not sought for the sake of the new, however historically unique modern natural science's preference for the new may have been. Rather, the new is continually to create innovations, which in turn expand the space for activity and possibilities. Audacity for an unforeseeable and fragile future is nonetheless needed. The expansion of the possibilities of and space for activity increases the complexity of events and thus the uncertainty of the outcome. The idea of susceptibility to planning, an idea that accompanied modernity from the beginning, has long since proven to be an illusory dream of controllability. But we still have to learn how to live the life of uncertainty that is inseparably tied to *innovatio*.

Research conducted by the sciences of complexity on chaos, self-organization, and networks has brought these concepts into everyday language and permitted them to serve the management of uncertainty. But everything that is thought and said about the future and all supposed knowledge about it ultimately depends on past experiences and already developed expectations. At the beginning of modern times, when the temporal horizon toward an open future began to expand for the first time, the tension between experience and expectation was especially powerful. The palpable discrepancy lay in the alienness of the expectations of the improvement of living conditions, of new knowledge, and of the technological possibilities that would contribute to their realization. Francis Bacon formulated these promises programmatically and the most trenchantly for the still young natural sciences. The certainty of salvation that came from the religious tradition and that linked experience and expectation with each other was first loosened and then exploded. Later came the "discovery of the malleability of society."[68] Today other parameters mark the field of tension between experience and expectation. Both have become extremely volatile because they hardly have time to congeal. The image of the future is no longer a static one; it changes dynamically in proportion to our attempts to imagine and conceive the future.

Past Future: A Look Back

Thirty years ago, the future was still regarded as foreseeable and certain. This, at any rate, is what we must conclude if we reread *The Limits to Growth*, the study that the Club of Rome commissioned in 1972. It was completed after just fifteen months of research work, sold more than 30 million copies worldwide, and was translated into more than thirty languages. What interests us here is not so much the reason for its unparalleled, strong resonance but the question of how this past future looks from today's standpoint.[69]

In the world model produced by Forrester, the Meadows, and their associates, the future was still considered highly foreseeable. Paradoxically, it was this belief in foreseeability, supported by the use of the first computer model created for this purpose, that secured a strong public effect for *The Limits to Growth*. The results of the study were accompanied by the galvanizing call to change human political and economic behavior in a way that would prevent the prognosticated collapse of the environmental system. By substituting probabilities calculated by the computer model for uncertainties, guarantees for the future were constructed to avert the development of crises. The gloomy predictions were accompanied by recommendations showing a way out of the crisis. The—unfavorable—expectations were more or less precisely formulated. The argument was that they resulted from prior experiences—namely, the irresponsible exploitation of nature, something that everyone could understand and that was evidenced by many local and global examples. The conclusion was that only a radical turning away from prior behavior could effect a turnaround.

Comparing this look back at a past future and the currently dominant image of the future leads to the astonishing realization of how much our conception has changed within a single generation. Today, speech about the future is in the subjunctive mode. The term *future* rightly ought to be used in the plural, even if our language resists. Uncertainty and contingencies, possible alternatives, wishes, and probabilities permeate the image and express themselves in many ways. The fear of the one great catastrophe—the environmental catastrophe—has been displaced by risks of various kinds and dimensions. Small and large risks affect private and public life. Scientific instruments, especially the simulation models, have developed further, and with them so have the extent and degree of differentiation of their statements. Their contents have become more provisional and fluctuate more intensely. They have become tools for thinking through the future in the sense of "what would be if"; they no longer claim to contain definitive statements.

Other quantitative and qualitative tools have developed for systematically dealing with the future. Their basic assumptions are more thoroughly thought through and are considered in the final conclusions. The spectrum ranges from back-casting to future-scanning, from retrospective historiography to the broadest possible spectrum of methods for creating "visions" and practicing foresight. Today, in dealing with the future, creativity is especially necessary. Flexibility, another characteristic of today's style of life, is transposed to life planning as well as to the instruments of planning. Ever since the victory of the market economy, the neoliberal credo, and the final collapse of communism, we have witnessed the global rise of belief in a mechanism that promises that it carries within itself a degree of adjustability and flexibility toward unexpected changes in the situation that is sufficient to deal with the unforeseeability of the future—the belief in the market.

On the level of research, one characteristic of many systems has come to command the respect it deserves—their complexity. In retrospect, the model developed at MIT by

Forrester, the Meadows, and their coworkers for the Club of Rome appears utterly naïve. Understanding of the phenomenon of complexity has grown apace with increasing knowledge, greater computer capacity, and the routine manner in which simulation models are put to versatile use in the science system. Finally, the behavior of many living systems that are subject to the laws of evolution is incalculable. In their evolutionary blindness, they are inherently stochastic, and with their behavior subject to selection, they are extremely nonlinear. They also resemble nonthermodynamic systems whose qualities are almost totally determined by statistical averages over a great number of almost identical states. The most interesting characteristics of evolutionary systems result, rather, from the dynamic magnification of extremely rare occurrences—for example, that for the first time *one* bacterium develops resistance to an antibiotic. These are not even singularities in the conventional mathematical sense for they cannot be reproduced in the language of functional analysis.[70] If we assume that human societies, too, display at least partially evolutionarily determined behavior, then it is clear how much complex systems differ from the mechanistic assumptions that still inhered in the model developed at MIT. From an evolutionary standpoint, our historically shaped thinking about the future moves in a border region between stability and flexibility, self-organization and the chaotic edge.

But in the face of this complexity, which was only barely perceived three decades ago, how is planning possible at all? Opinion still differs on this question. For the postwar generation that experienced the "golden three decades" up into the 1970s as the blossoming of modernity and that strove to do everything possible to catch up with the United States in terms of societal modernity, much seemed amenable to plan and feasible that today, considering the loss of the state's central control capacity, would seem doomed to failure. In a time when "governance" has replaced "governing," which is now rejected as too statist, we look back with astonishment at the political measures and means of control that were developed at the beginning of the 1970s. But what seems so alien on renewed reading of the Club of Rome study is not just the technocratic aspect that was clearly present in the first models but the belief in its political practicality and realizability, which was presented with conviction. It is the *lack of another perspective*—the lack of a view from below, from and of the local levels, the lack of the inclusion (now at least rhetorically taken for granted) of "imagined lay persons"[71] as consumers, voters, and users—that seems like a gap today. One asks in astonishment how it is possible to speak of the future without listening to the people it will affect.

Perhaps we really have lost our future—in that we have lost the feeling of being able to control it. The sociologist Anthony Giddens says that it is this loss of control that turns the future into a problematical category.[72] Suffused with thinking in concepts like competition, risk, security, and globalization, planning becomes more, not less, indeterminate. The future has ceased to be a space that can be conquered and colonized, as the Enlightenment still thought possible. The future has become a solid component of a modernity that cannot stop being modern.[73]

Thus, people have once again lost their illusions and have suffered one more blow to their narcissism. Even the futurologists, the professional agents of this colonization, have realized that the conventional approach to planning the future no longer functions. If in our perception the future has become less foreseeable, then this is partly because of the expansion of scientific-technological knowledge, which is accompanied by new uncertainties because it expands the number of possibilities and demands new decisions. The greater heterogeneity of knowledge and the expansion of the group of stakeholders (the various social groups whose opinion, behavior, interests, and values must be taken into account) are increasing the degree of complexity. This makes possible a reflective turnaround in our thinking about the future. The point is no longer to "predict" the future, if such a claim could ever have been taken seriously. Nor is the point primarily to show the trends and possibilities of developments. This remains an attractive business[74] and one that remains fascinating if carried out with intelligence. It stimulates the imagination and allows us to entertain alternative standpoints. But the questions have fundamentally shifted.

What has become more important today are the questions about how the various actors imagine the future, how they approach it, and what time frames they construct for their various purposes and interests. The approach to a large number of subjective viewpoints is part of a broader picture that conveys complexity and calls to increase awareness of other actors' viewpoints and options for action to take them into account in one's own deliberations. The way to even an only approximately objective picture leads through a multiplicity of subjective viewpoints and sites. It has also become clear that every publicly conducted discourse on the future takes on important social functions that cannot be determined in advance. Discourses *on* the future therefore often try to organize a kind of competition *over* the future. The rhetoric that the various agendas for innovation and for the future make use of and the target groups that are to be addressed aim to empower the latter and overshadow all statements about what "the future will bring." Future-oriented political programs serve the attempt to occupy certain areas of politics, and various predictive toolkits (as they are used, for example, in foresight programs) stand for the effort to use political instruments to create a shared future.

But what is the situation with today's crises? Has humanity lost its consciousness of crises? Has this consciousness given way to an unsurveyable realm of uncertainty and subjectively experienced insecurity? The appearance of global terrorism has made these questions intensely urgent, but they are not really new. Many of the problems addressed in *The Limits to Growth* remain unsolved to this day. With other problems, the emphasis has changed, and new problems have arisen. In regard to the environment, attention no longer focuses on the exhaustion of natural resources or primarily on the problem of pollution. The topic dominating all others is rather climate change, which stands for the unforeseeable per se—for extreme fluctuations in weather and climate, for extremely varied effects with a local and regional scope. It stands for a feeling of self-induced, anthropogenic helplessness whose only prognosis is to expect the unexpected.

In the age of globalization, global consciousness has also increased. The world is increasingly perceived as a whole while indifference toward local phenomena has grown. In a certain sense, inequality blinds people to the suffering and unsatisfied desires of others, especially when the population of the world continues to grow, though at a slower pace. Humankind's crisis situation that the Club of Rome warned about—prophetically, ideologically, and, as critics have noted, without considering the economic laws of the market—has eased, but the crisislike manifestations and effects have spread over the entire globe, and new ones have been added. They have grown locally more visible and globally more invisible. The one crisis overshadowing all the others (which, revealingly, was not even mentioned in *The Limits to Growth*) was the real danger that an annihilating nuclear strike between the two superpowers would extinguish the human species. Today, this threat appears in the decentralized form of uncontrolled nuclear proliferation. It no longer stands under the sign of the conflict between the superpowers but under that of the so far hardly inhibited spread of a globally active terrorism.

Today the one great catastrophe has been replaced by manifold risks whose causes, appearances, and effects constantly change, and a past, imagined future has turned into an actually uncertain present. The past future appeared catastrophic, but it was perceived as certain. One could believe in it—or not. One could act to prevent the supposed catastrophe. But today, the uncertain future calls for strategies of action that cannot be narrowed down. The future we now face relies on innovation under conditions of uncertainty. This cannot be equated with lack of knowledge—quite the contrary. Uncertainty arises from the surfeit of knowledge, leading to too many alternatives, too many possible ramifications and consequences, to be easily judged.

And yet not everything in this complex process is opaque and unforeseeable. Here, too, human action requires normative conventions and rules, a common foundation of rights and obligations that gives meaning to practices and that can demand mutual accountability. Innovation, as the collective wager on an uncertain future, thus includes the possibility of a misstep or failure. Things can tip over at any time, and every vision of the future can turn into its dystopian opposite. With this possibility of failure and of tipping over, which must be seen and consciously accepted, modernity returns to its original premises.

Modernity's Promises

The idea we currently have of the future is comparable to a probe connected to a spaceship by a tether. It was sent out to carry out measurements in an unknown and uninviting environment. The probe is anchored to a tiny base, while the space it is to explore is potentially infinite, cold, and indifferent. The probe continues its explorations, unimpressed and tied to the mother ship, driven by human imagination and insatiable curiosity. It feels its way forward along a time axis whose scale cannot be precisely read. The probe is equipped with organized determination. Human imagination serves it as an aid to orientation—vague promises of a wide variety of improvements; the desire to understand, which accompanies the will to control; but also the unbroken joy in playing with the next big thing that does not know its goal or purpose. There are no historical precursors for this way of dealing with the new, which raises innovation to the status of the definitive though indefinable center of the realm of the future that is to be explored. What is coming, the time and space of what is not-yet, is perceived as something overflowing with constantly generated opportunities and chances. Like information, which is regarded as the new raw material of the electronic age, chances are present in plenty, but they must be selected, processed, and combined with other chances, data, and networks. Opportunities are there to be grasped, but they have not yet been cashed in. They remain promises.

That is the collective wager we have made on the future of scientific-technological civilization: it is called innovation. The social order of the highly industrialized countries is to carry out the promises of the order of knowledge and things as the primary goal on the basis of economic competition with economic growth. In an inescapable pincer movement that is flexibly arranged in accordance with the neoliberal economic dogma, we approach the horizon of an uncertain future. No one has reason to draw back from it, we are assured. Uncertainty is inseparably a part of it. If the knowledge of contingency stood at the beginning of modernity, we are currently in the process of discovering uncertainty as an inherent component of the process of innovation and a fragile future.

Where does innovation's incontestable pioneer role come from? Creating a connection to an unknown future has always fascinated people. Every culture, every historical epoch strove to read out interpretations or instructions for action for the present. The wish to make predictions promoted ancient ideas of numbers and mathematics in China. Christian theology sketched a picture of the future that was bedded in the idea of salvation. But in many societies, most of history was dominated by the overarching idea of fate, and fate regulated the life of most people. Only in modernity did the idea arise that the future could be planned and current living conditions bettered, at least in part. The steps of progress that were made in various areas of knowledge—for example, the mathematical theory of probability—made it possible to provide economically sustainable foundations for newly arisen institutions, and today the insurance industry is generally considered an acceptable way to deal with the uncertainties of the future.

Modernity's promises were also based on new confidence in the achievements that, after "the discovery of society," were at least partially carried out in increasingly self-organized and self-regulating form. The belief in the possibility of planning and carrying out technocratic ideas that was still widespread a few decades ago has dwindled today to the point of recognizing limits and being ready, at least in principle, to cater to the wishes and expectations of groups like consumers and voters. Today, innovation has become the starting point for negotiating with the future. It takes account of the future's uncertainties and emphasizes the opportunities opening up. The concept of innovation has changed, as well. While the new configuration of known elements and components is still the core of the concept, it has expanded in a direction that allows the radically new in a socially evolutionary sense.

But why precisely now? Where does the collective obsession with a continuous process of innovation—a process no longer fixated solely on producing the new but also demanding its productive, profit-making implementation with increasing returns and its embedding in social and economic living circumstances—come from? Even basic research—that segment of the production of new knowledge and discoveries in which breakthroughs press back the boundaries of what is considered feasible and that provides indispensable epistemic foundations whose significance is often not visible until later—has not slowed its approach to possible technical applications and expectations of innovation. Most researchers in this area are indeed aware that behind the walls of the laboratory, behind the fascination of their experimental systems or research apparatus, and behind the computer screens on which they create models, expectations are gathering in accordance with which they "translate" their knowledge and bring it into a form that, sooner rather than later, should be in some way utilizable by and indeed useful to the society. Curiosity and the desire to discover have not disappeared, but they have lost some of their former independence and self-understanding. They have become part of the research machinery that is expected to point the way to innovation.

The current focus on innovation does not mean that this is a new phenomenon or that earlier times did not already equally look on innovation as desirable and central to economic growth. But neoclassical economics still has trouble making sense of innovative processes and Schumpeter's approaches to a theory of innovation.[75] Economists who seize on Schumpeter for modeling or empirical investigations of processes of innovation ineluctably move away from neoclassical theory. From an empirical standpoint, processes of innovation are the result of specific activities that aim at introducing new products or at altering production processes. At any rate, they cannot be understood as something routine, whose results are foreseeable in detail. The theoretical problematic lies in the fact that the success of innovative procedures is unforeseeable and in the resulting problem of discontinuous change. Investments in innovations cannot be rationally calculated because they face the strategic uncertainty resulting from the actions of others and the uncertainty of the innovation's usefulness. As Joseph A. Schumpeter already argued long ago in *The Theory of Economic Development*, entrepreneurs are interested in profit, but innovation is not exhausted in the motivation of a goal-oriented maximization of use.[76]

Conceptual Empty Spaces in Thinking about the Future

Language is sometimes able to assimilate the new that is already present but not yet named. The term makes the new comprehensible until it becomes taken for granted. In the nineteenth century in the United States, the term *technology* first emerged and contained a response to two long-term developments that had first become noticeable in the 1840s. One of them, according to the historian of technology Leo Marx,[77] belonged in the history of ideas. It consisted of ideas about what the "useful" or "mechanical arts" were and what purpose they served. The other development surveyed the organizational framework and created within it space for the concrete technical products and work methods of the incipient industrial age. The history-of-ideas development subtly but profoundly changed the ideas accepted until then about the relationship between technological progress and society's belief in progress. The Enlightenment had essentially still seen scientific-technological progress as a means to achieve socially and politically desirable goals. The term *technology*, however, made it clear that the supposed means had in reality become an end.

In a public speech delivered at the opening of a new section of the Northern Railroad in New Hampshire, Senator Daniel Webster praised the "extraordinary era in which we live" and paid homage to the "progress of the age" that "has almost outstripped human belief; the future is known only to Omniscience." The railroad had ceased being a means of populating the country or tying vastly distant parts of it together. The young, upwardly striving stratum of entrepreneurs comprehended for the first time what the "mechanical arts" stood for: they were concrete objects, infrastructures, and projects. They were the railroad. These changes in the way of viewing things, said Leo Marx, were what introduced the new term that gave expression to them. The mechanical arts were replaced by the new concept of technology.

The second long-term development unfolded in the organization of the useful arts and their material content, the machines. The individual machine was replaced by a sociotechnological system that introduced new orders of magnitude and incisively changed the organizational preconditions. The railway connections created the earliest, most visible, and most extensive major technological system of that time. In this system, the mechanical components were still indispensable, but they were now part of an expanded whole. At the same time, the organizational demands on its functioning increased, beginning with auxiliary equipment and deliveries and including the enormously increased need for capital for investments, which only the newly emerging great industrial and trading companies could raise. The demands placed on labor's skills changed in equal measure. The coalescing of scientific foundations with the "practical arts" and industry was already underway in many practical areas. But the entire extent of the transformation did not become visible until the end of the century, when the new system, in particular the electrical and chemical industries, reached the zenith of their growth.*

To understand why the process of innovation has become a central concept of the present, we must go beyond economic considerations. Innovation occupies a conceptual vacuum in our collective imagination of the future. This is an important empty space that promises a key for a future that cannot otherwise be found. All thinking about the future is itself historically determined. Our collective imagination has shifted from the belief that the future can be constructed and planned to a heightened sense of its unforeseeability due to its complexity.

* As so often in history, here too there was a pioneer predecessor. In 1829, the botanist and physician Jacob Bigelow of Boston suggested the term *technology*, a word he said was "sufficiently expressive" to include the practical application of science that "may be considered useful, by promoting the benefit of society, together with the emolument of those who pursue them." But the greatest success in the word's spread came when it was taken up in the name of an institution, which has carried it since 1861—the Massachusetts Institute of Technology, or MIT. See note 77.

From an evolutionary perspective, this image of the future looks out onto a radical openness. The projections therefore oscillate between the idea of an emerging order that increasingly arises through self-organization and the feeling of standing on the edge of an abyss. The oscillation takes place in a precarious balance between a suitable degree of stability and a fundamental openness toward the unforeseeable, including the singular events it involves. With increasing knowledge about complex systems and their dynamics, thinking about the future has grown less mechanistic and naïve. Perhaps it has become more reflective in that questions like "What will probably happen?" no longer have primacy. The questions have shifted to the knowledge and imagination of the various social actors and attentively follow how they construct their ideas of the future. Particularly in one area, the financial markets, this way of posing questions with the aid of mathematical tools has reached a high degree of sophisticated elaboration.

Imaginary constructs of the future fulfill various social functions in public and private discourses. But they have also entered into the various innovation scenarios with the intention of mobilizing the cultural, economic, and social resources that are considered the indispensable prerequisites of technical innovation, particularly from arousing desires and articulating latent or manifest needs through the production of identification with new products to the decision to buy or invest on the market, in production, or in the service sector. Public discourses on innovation, the rhetoric that is thereby used, and the target groups addressed have become at least as important as the conveyance of concrete contents. Discourses on technological pulls or pushes, the impetus and consequences resulting from innovations, are based on the assumption that new technological developments are accompanied by corresponding demand and social acceptance. This—mechanistic—viewpoint has lost credibility in a society whose recognized plurality today demands that viewpoints, interests, and voting decisions take the various stakeholder groups into consideration. While the expansion of the group that is regarded as relevant actors and their viewpoint on the future necessarily increases the degree of uncertainty, it nonetheless promises to expand the number and kind of possibilities that present themselves.

There are many reasons for the emergence of a conceptual void in thinking about the future. One of them is the changed relationship between the market and the state. Because innovation is a sociological process as well as a process resulting from the knowledge produced by science and technology, the promotion of innovation activity and of entrepreneurship is increasingly in the proactive purview of governments. In their influential 1985 study *How the West Grew Rich,* Nathan Rosenberg and L. E. Birzell Jr. noted that in well-ordered societies, political authority is dedicated to stability, security, and the status quo, making it singularly ill-qualified to channel activity intended to produce instability, insecurity, and change. One of the things this referred to was the state's promotion of innovation.[78] Today, all highly developed industrial states have a bundle of policies that aim to promote investments and scientific-technological innovation. Although technological innovation in the narrower sense is still carried out primarily within private companies and can thus be defined as "the successful implementation (in trade or management) of a technologically new idea for the institution" or as "the process by which companies master product design and production processes that are new to them and introduce them in practice," today it is widely acknowledged that technological innovations need a broader, innovation-friendly environment if they are to prevail. Ultimately, an innovation-friendly society is sought.

A political agenda that aims to promote scientific-technological innovations represents the production of a common idea of the future by means of a political toolkit. This process reveals the necessity of both actively and interactively dealing with the fragmentation and indecision of collective activity in relation to the future. It is precisely *not* a matter of course for processes of innovation to follow from new knowledge gained in basic research, however great the innovative potential of this knowledge may be. The "linear model" according to which there is a preset path transporting new insights and results in basic research to applied research, which in turn someday bears fruit on the market in the form of new products or production methods, is merely an idealized version of a historical process that predominated after the end of World War II. Nor can processes of innovation be left to companies alone, however impressive their creative "unrest" (in Schumpeter's sense) may be.

The preconditions can be listed, but they are no guarantee of success. Among them are social and human capital—that is, well-trained people and the networks and suitable forms of organization that aim to connect them with each other under adequate conditions of competition and cooperation. The other form of capital is also needed, particularly venture capital. Individual resources are needed, like the imaginativeness and the creativity of successfully assertive individuals and groups and the flexible institutions and state regulating processes that must foster them. Too much planning, too much regulation, and too much centralization inhibit not only research but also the innovative potency.[79] Other barriers to innovation include a reticence to engage in risk and to make creative mistakes. There must also be a knowledge basis that wants to expand and a research system whose orientation roughly corresponds to society's expectations since broad public approval and support can be decisive for specific directions of research.

The more effective a renewal is, the greater the changes that appear—for the gain of some and the loss of others. As in earlier waves of modernization, there will continuously (and unfortunately also cumulatively) be losers and winners. This is why innovation always encounters resistance and rejection. They do not always appear openly but can conceal themselves inside institutions and in the inertia inherent in large systems. Nor do innovations necessarily always provide the best technological solutions. Technologies can become locked on the same level. When new technologies are introduced, there is a strategy for consciously bringing about this situation to increase costs for competitors.[80] All this (and more) is generally known and is part of the public discourse, whose intention is to push society forward on the path to a fragile future. The harder it seems to grasp the goal, the more important is the path to it.

Innovation Fills a Gap

The Enlightenment's dream that progress in the sciences and in technology would be the instrument for social and possibly even for moral improvements and would lead to political emancipation in the sense of an overarching rationality was short-lived. Science was in reality unable to free people of their passions—or it could do so but only by means of interventions in their neurochemical system, as depicted in *A Clockwork Orange*. Scientific and technological achievements have not prevented repeated regressions into barbarism, of which the twentieth century had its excessive share. Indisputable improvements in the quality of life and living standards and even in the standards by which people live with each other, as expressed in human rights, can be countered with a long list of intended and unintended side effects that are directly or indirectly tied to science and technology.

The risks that have spread due to interventions in the natural environment are firmly anchored in present-day consciousness. The shock waves triggered by books like *The Limits to Growth* or Rachel Carson's *Silent Spring* in the late 1960s and early 1970s led to a changed consciousness of our dealings with nature. The risk society became omnipresent. Risks are encountered in the area of health and nutrition, they threaten desired social security, and they escalate in official alarm levels to a bundle of old and new dangers, currently focused on terrorism. When Albert Camus wrote of the twentieth century as the century of fear, he had in mind the terrors of totalitarian regimes and the growing arsenals whose destructive capacity was enough to eliminate much of humankind. In the light of recent events, this fear has not diminished; only its profile has changed.

The fear in the foreground is no longer of a great catastrophe. Fear comes in smaller doses whose effect is all the more lasting. Many of the actual or potential risks that trigger fear remain invisible, and their consequences are long-term. Unfavorable consequences that arrive late overshadow the present, although the present can claim to have increased chances for a healthy, longer, and better life. The suspicion that many of the newest scientific-technological breakthroughs bear an invisible risk that is not yet recognizable but that will appear later all the more virulently does not seem destined to dissipate from the public's consciousness. The ideological vacuum that has arisen after the disappearance of an often naive belief in progress has been overshadowed by the suspicion of the riskiness of all scientific-technological innovations—and yet it waits to be filled by something positive.

The great sociotechnological systems that were the trademark and pride of modernism have not disappeared. Their unsurveyability manifests itself as complexity, which can bring about "normal accidents"[81] but which is—still—regarded as controllable, in principle. But the vulnerability of these systems meanwhile cannot be overlooked.[82] The great spread of information and communication technologies[83] has led to decentralization. The creation of economic value is being brought closer to the user as it is distributed spatially or shifted to sites where the availability of qualified but cheap workers is relatively high. The world of the factory, which was once characterized by planning and control, hierarchical structures, and the processing of mass commodities, has in part made way for a high-tech world based on the processing of information and knowledge. The hierarchies have flattened, and the technologies used stand in close relationship to other technologies and highly technologized products. Work itself, in the form of projects, is carried out by heterogeneously composed teams whose abilities and skills supplement each other in a highly qualified way. Activities that were once carried out directly by humans are now processed indirectly by software. Adjustment to a constantly changing environment is considered *the* guideline.[84]

With the shift toward market forces, neither the jurisdiction nor the extent of state regulation has substantially diminished. What have become more permeable are the boundaries between states. Workplaces are outsourced, and incentives for investments are created elsewhere. Whereas the usual practices of management and engineers in the period between the world wars was greatly influenced by the managerial style of large corporations, after World War II, as Thomas Hughes shows, engineers and managers were confronted with the task of introducing new technological systems like computer networks and city highways. These systems are much more heterogeneous in their composition; they tolerate heterogeneity and engender it themselves. The expectation of discontinuous change is built into them. Their daily management is based on the principle of discontinuous complexity. Where "modern" project and technology management was based on hierarchical and centralized control mechanisms that found their material correspondence in tightly coupled systems, standardization, and homogeneity, the "postmodern" regime relies on flat, horizontal, networked control mechanisms that correspond to loosely coupled and heterogeneous systems. Control, we are reassured, functions in this technological culture without needing a nerve center. The disorderliness of complexity is accepted as part of the bargain to steer development in one direction or the other, depending on the signals coming from the market.[85]

The expression *technosciences*, which increasingly appeared in the 1980s, is often used to signal the close connection that science and technology have entered into. Many scientific discoveries are pushed forward by new technological instrumentation, which in turn is the result of the production of knowledge in the context of specific technological problems or problems presented by users. The epistemic and technological things have found each other; they have become knowledge-technological objects. One result is that the sites where new knowledge is produced spread rapidly and multiply via the technological infrastructures, including research infrastructures. New research instruments and technologies and technologically supported methods can spread faster through the various disciplinary fields. Interdisciplinarity arises due to the pressure to apply or jointly develop technical infrastructures and methods in other disciplines. Transverse technologies create new practices and fields of practice across all existing disciplinary boundaries.

In the same way, the exchange between the laboratory, industry, and the service sector is intensifying. Released from the laboratory, the knowledge-technological objects enter a social environment that is itself technologically equipped and scientifically well-developed. In this way, according to the ideal, use is tied anew to users and new uses. The thus enriched potential for use can continue to have effects. The market ensures that the knowledge-technological objects are brought into the required flexible form tailored to specific purposes and desires. They are miniaturized, user-oriented, interactive, and much more so that they can fit into numerous, widely distributed networks that are heterogeneous, disordered, and complex. The connection between "knowledge what" (propositional knowledge) and "knowledge how" (prescriptive knowledge), referred to by economic historian Joel Moykr,[86] has become extremely potent.

Other changes that affect the management of the spatial and temporal coordinates of knowledge-technological objects are the result of a shift from exotechnologies to endotechnologies. Technology has developed since the early history of humankind, independent of (proto-) scientific observations, theoretical speculations, and the search for systematic patterns of explanation. Technology offered protection to people living in social groups and gave them growing control over their environment, in which they learned to process and extend the tiny ecological niche they began with. Technology assumed the function that archaeologists and anthropologists have attributed to it—to compensate for and, step by step, overcome the biological limitations of *Homo sapiens*. Exotechnologies aim at the expansion of possibilities of controlling the environment. They have enabled people to travel greater differences in less time and to settle the space they found more densely and efficiently. The processing of found and extracted materials finally enabled the mass production of artifacts, the preservation of foodstuffs, and the erection of infrastructures that in turn made it possible to live comfortably in otherwise inclement climate zones.

In contrast, the regime of the endotechnologies—bio-, nano-, info-, and other converging technologies—changes the dimensions and scope of action of the scientific objects. They form mostly invisible yet visualizable infrastructures that can penetrate into the smallest dimensions of matter or living organisms. Via the genetic regulatory mechanisms, on the one hand, and variations in the speed of information transmission, on the other, they change the rhythm and management of time. Natural aging processes can be slowed or sped up; the flowering of plants can be retarded or reversed. If the supply of electric current, an exotechnology, once extended the day far into the night and thus incisively changed societal life, now endotechnologies can intervene in the circadian rhythms of living beings by switching genes on and off. The molecular and cellular interior of living organisms is becoming a realm in which targeted interventions can be made. Only the vaguest contours of the possibilities and consequences for the process of the origin of life, growth, aging, and decline on the various levels of organisms are thereby beginning to emerge.

In December 1959, Richard Feynman presented his now classic lecture, "There's Plenty of Room at the Bottom," before the American Physical Society. In it, he addressed the possibilities of manipulating and controlling matter on the tiniest level. He foresaw that the interior of matter would become the primary site of the development of knowledge-technological objects and endotechnological procedures. On this level, which is the smallest currently accessible to interventions, atoms can be put together and manipulated at will. Today, artificial environments are created in which various surfaces are brought together to introduce completely new sensory, organic and inorganic, endocrinological, or neurological connections and to make exchanges between them. The growing inter- and transdisciplinary collaboration between biology, mathematics, physics, chemistry, computer science, statistics, and other areas of knowledge is beginning to converge in a common research agenda. The life sciences, among others, are striving for an integration from the molecular level upward that will include the organism.

What the scientific community has received with enthusiasm is creating unrest in the public realm. New questions are raised—for example, what being human will mean in the future. The possibilities of endotechnologies surpass all the promises modernity ever made—although the desires on which the promises are based have never been clearly articulated. The most recent wave of renewal did not come in isolation but forms a greater cultural pattern. The more radical the renewals are from a scientific-technological viewpoint, the higher the proportion of social knowledge must be if society is to be put in a position to appropriate them culturally and thus transform them in a way that gives them sense and meaning. The renewals require a language because everything they have brought so far is the sober, technological jargon of their specialist producers or the colorful, multimedia images of public relations and marketing. They have to be described—in a way that makes them relevant for the living contexts and images of the future of those who are to use them. In other words, they must become culture themselves.

The Ambivalent Answer of Innovations

Here the term *innovation* enters the existing vacuum and begins filling the gap that discontent left in society. Innovation signals the emergence, the arising of something that may already be present but is only partially recognized or recognizable. It directs the gaze to invisible or unexpressed connections to other terms and weaves them into a net of newly configured meanings.* Despite all attempts to celebrate postmodernism as an epoch that has broken with modernity, we retain modernity. We are condemned to modernity and probably also for a long time to the need to confirm it anew and to give it content, even if the design, the irony, and the reflectivity with which content is conveyed, altered, lost, or shifted become themselves components of content.

Modernity is no longer the program that provides answers to prefabricated utopias, even if the building blocks that it once contributed to furnish institutions and societal structures are still present and usable. Nor does modernity answer any longer to explicit or implicit expectations: the majority of its promises have been fulfilled, even if differently from what was expected. It still functions as a substitute for a belief in progress that has collapsed under its own hubris and the illusions it created for itself. But modernity must continue to create itself out of itself,

* In the 1960s, when the cultural scholar Raymond Williams investigated the changes that the term *culture* had gone through in the first half of the nineteenth century in England, he discovered a fascinating correlation between societal change and change in the language. The term *culture* stood in close connection to other key terms of the time (like *class*, *industry*, and *democracy*), and it altered its meaning in response to the societal changes that were expressed in the other key terms. Raymond Williams, *Culture and Society 1780–1950* (London: Chatto & Windus, 1958).

and to this end a future is needed that is radically open and uncertain. It is accompanied by the belief and the ideology that something new can be created that bears within it the effective power to create additional valuable new things that can contribute to societal affluence and well-being—however vaguely defined these may be.

The process of the generation of the new must be adequately open for societal values—for example, as they are expressed in the demand for sustainability. Other values, like security, which presents itself in a number of mutually contradictory meanings, also wait to be included in the next package of modernity. In contrast to the term *evolution*, aside from the problematic of the term's metaphorical transfer from biology to the area of living together in society, the term *innovation* holds room enough for societal values and human action.

Innovation proves to be a slippery term that is vague enough to remain in flux. Thus, it can easily shift terrain because innovation is in demand and potentially useful everywhere. Demand exists in the realm of culture to make connections with the creative industries, to mix the genres creatively, and to create a system that is itself in turn fed by the innovative power of subcultures that make use of the new technologies, of design, and of wittiness. It is needed in organizations, where managers want to institutionalize it for the long term to create the flexibility that companies need to adjust to a changing environment but also to create a favorable environment.

Because innovation is open for human action, it can endure mistakes. It can reassure people to keep trying. It indicates failure as well as the successes that can be taken as examples and displayed for it knows only one form of proof—success. The path to success is arduous, thick with setbacks and uncertainties that should not be swept under the rug because innovation seeks to encourage and create the audacity that is needed to enter into skillful negotiation with the unforeseeable and fragile future.

The initially abstract-seeming goals of the next wave of innovation—for example, in the life sciences (or should the term be *life technologies*?), which correspond to the desire for health, longer life, the postponement of aging, or expanded possibilities in human reproduction—can be achieved only by means of research processes whose consequences and prerequisites are far from understood. But the discontent with biotechnology and related research directions, despite the attempts to underscore their directly useful, therapeutic purposes, has deeper roots. What manifests itself here is the feeling of ambivalence. On the one hand, this ambivalence greets the improvements that are made to seem possible. It plays with our inexhaustible wishes and fantasies. On the other hand, it stands as a direct threat to the seemingly immovable idea of identity and self, of kinship relations and other socially or culturally molded networks of relationships. And yet these are not given by nature but made by us humans. To carry out this mental step, however problematical it may be in a concrete situation, also requires something of the audacity, the heroic gesture, that innovation demands in full knowledge of the possibility of failure.

No wonder that behind every supposed or real shift in the boundaries between the self and the others, behind every intervention in allegedly immovable natural limits, and behind every discussion of an ethical case, we hear the invocation of unchangeable values and an ethic that offer guidelines for a firm orientation in the middle of the unhinged categories. But numerous historical examples evidence* that not every slippery slope necessarily leads into the abyss. Values may appear unchangeable, but they are subject to societal changes against which they cannot muster enough immunity. Caught up in the tension between the understandable wish to preserve the given order, thereby resisting change, and a new hybrid order whose contours are only vaguely visible, the impulse finally prevails to move forward because standing still is tantamount to falling back. This is another, deeper reason why we are condemned to modernity. Innovation is the social form that this impulse takes on to move toward the new order.

The meaning of the term *innovation* changes through the process that the innovations go through. Innovation no longer stands primarily for a recombination of known components or elements, as Joseph Schumpeter presented in his classical analysis at the beginning of the twentieth century, which gave Schumpeter's idealized figure of the entrepreneur a crucial

* A little-known example is the controversy that arose in the late nineteenth century in the United States about contracting life insurance. The opponents rejected it vehemently with the argument that it would be immoral for human beings to presume to make the time of death, which God alone could know, into the object of a business deal. The controversy gradually ended only when the counterargument was fielded that taking precautions for the children and widow of the deceased was also a moral duty.

advantage over his competitors. This form of innovative behavior is as widespread as ever. But it has been supplemented by an expanded concept of innovation that builds on the potential of the radically new and the associated inherent insecurity. In the 1970s, the economist G.L.S. Shackle already called the "essentially new" the characteristic trait of an evolutionary approach to investigating sociotechnological innovations that are radically open in the face of an unknown future.[87] The process of innovation presumes contingencies and decision-making situations that surpass by far any recombination of known components.

In this way, innovation fills the vacuum arising from genuine insecurity, which in turn is an indispensable part of the process of innovation. Innovation can fill this vacuum in the interpretation of the future because of this circularity, which in turn is a form of modern reflectivity. Innovation is not an unmoved mover that acts by means of the impersonal powers of a technocratically organized society. Technocracy itself, as a historically molded societal structure, is undermined by the process of continuing innovation in the form that includes the radically new. The process of innovation spreads throughout society, just as its scientific-technological objects and their network-organized infrastructures do. A kind of mimicry is at work here, but it is not exhausted in mere imitation. It imitates by actively intervening and appropriating what it can use. It imitates, but at the same time it interprets and commentates and changes the meaning by inventing new meaning. There is no longer the one great actor, either Schumpeter's figure of the entrepreneur or the state and its agencies. Now all are called on and empowered to take part in the process of innovation—at their own risk, of course!

In this way, innovation presents itself as a key concept and as the only currently available and plausible answer in the face of the insecurity to which it constantly contributes, aside from the various kinds of fundamentalisms that offer their deceptive securities as an alternative. Innovation's credibility comes from the basic willingness to include the diversity and plurality present in the society. Variation is the precondition; it is cultivated to make selection possible because the process of innovation is based on both. Diversity also means that a scientific-technological innovation cannot be successful without social and cultural innovations. The more sweeping and radical the scientific-technological objects or technologies are, the more cultural and social innovation is needed if the package containing the renewal is to be embedded in the societal context.

A language is required if something is to become recognizable and pictures are required to convey what language cannot or cannot yet express. Literature and the arts are invited to pose questions, express doubts, and make ironical commentary. What initially narrows scientific-technological innovation and brings it into a rigorous form must then be culturally expanded again to do justice to the plurality of society and its future users but also to take their ambivalences into account. In this way, innovation becomes a key that seems able to open all doors, whether to escape the threats of the present or to dare a leap into the unknown future. The insecurity inherent in innovation corresponds to the openness of the future. Innovation cannot anticipate anything; it seeks to include, not to exclude, even though its nature is that it will exclude.

Here we are not yet speaking of innovation's forms of expression and its consequences. They already manifest themselves in preliminary form—for example, in the artistic, creative professions. For people who work in these fields and who regard themselves as innovative and creative, this means repeatedly and voluntarily letting themselves in for the outstandingly staged forms of competition whose function is to maximize variation to carry out selection all the more rigorously. This proceeds under the sign of seeming transparence and in adherence to the criteria shared by all, though these can be unjust and disputable in individual cases.[88] The mixture of the various genres and the exploration of new technologies are part of the creative way of dealing with them. At the same time, the ineluctable logic of economics enters into every creative act.[89]

Innovation offers enough room for human action and does not hide its dependence on such action. Where else should innovation come from if not from human activity and, of course, the preconditions that make innovative activity possible? These are the forms in which cooperation and competition are regulated, the social skills as well as the ability to manage the creativity of others, venture capital that is available to small start-up firms, but also the framework conditions and the rulebook governing the behavior of the collective actors, transnational corporations, governments, and interstate institutions. Science and technology's offer to produce the new is always an oversupply. The result then depends on the continuing course of the process of innovation, on a product's successful placement on the market, and on the goals of the productivity spiral, which can be achieved in many different ways.

The term *innovation* plays with the multiple meanings inherent in changes that know no precise goals, since these too must remain open. By betting on human action, innovation should broadcast the calming message that the unforeseeable will nonetheless be manageable. Dealing with risks? Not a real problem, for if needed, there is the precautionary principle. Instead of being intimidated by the apocalyptic warnings of the risk society, the decision can be made in favor of a "modern risk culture," as prevails in the global financial markets. These are institutions that cultivate a culture of risk in which market configurations and technological developments come together within certain social groups that develop their own ideas of leadership, expertise, and creativity. There the ambition to know encounters the market's ambition to turn this knowledge into commodities.[90] That too is a form of innovation—though certainly not for all.

In contrast to other, related terms (like *knowledge society*, which evokes counterterms like *ignorance*), the term *innovation* captures the essence of modernity in its iterative dynamic. This expects fractures and continuities, successes and failures. For innovation contains a self-fulfilling prophecy—namely, that only further innovations will provide the means to master the problems that innovation also creates. This circularity is solidified by the proof adduced by innovative achievements and in this way opens the present for a fragile future.

Epilogue | Why We Must Remain Modern

We live for tomorrow without enjoying it today.

—Casper Thyksier, at An Interdisciplinary Workshop in Cambridge, UK, September 2004

The first mathematical models that were applied to the financial markets were crude models of what really happens. They were of little direct use for actual investors because they were too weak to make predictions. Gradually, the models grew more sophisticated, and as a result they were increasingly used in everyday business. Word got around that the complex processes of stock market trading could be captured in mathematical formulas. One of these formulas—the one that Fischer Black, Myron Scholes, and Robert C. Merton invented in 1973 for the pricing structure on the option market*—was an important breakthrough in modern mathematical financial theory. The

* *Option:* The contractual right to buy or the contractual right to sell up to a specified amount of a designated security or commodity at a specified price and time.

following years showed an increasing correlation between the mathematical model and the observed prices because traders had begun to apply the model to identify and use for their own purposes the discrepancies between the model's predictions and market prices, thereby reducing the discrepancy. The traders began using the mathematical formula as a tool for influencing the behavior of other market participants and thereby the social reality of the process. The formula developed a performative effect. It taught its users how to price derivatives and how to improve their risk distribution.[91]

It is hard to imagine how the global financial market in derivatives could have grown from zero in the year 1970 to a total notional 134.7 trillion dollars in 2001 (as cited by MacKenzie) without the application of mathematics. Robert C. Merton, who was coawarded the Nobel Prize for economics (officially the Sveriges Riksbank Prize in Economic Sciences in Memory of Alfred Nobel) and who, as the son of the sociologist Robert K. Merton, may be sensitized to the dialectic between social reality and knowledge *of* social reality, remarked on this that "reality will eventually imitate theory."[92] This is one of many examples of how an economic theory not only describes real processes but is also capable of creating them. Economics—and not only economics—is a performative activity.

Could the omnipresent invocation of innovation—but with the crucial difference that there is currently no theory of innovation—have similar effects? Could it become a self-fulfilling prophecy that ultimately sweeps everyone along with it and even give rise to an ideal like *sustainable innovation*?[93] Why did this collective obsession with a term and with what it suggests arise precisely now? Part of the answer is anticipated in the self-evidence concealed in the question. What else but an increase in economic growth, a further turn of the spiral of productivity, and a continuous discovery of latent desires and social uses, of new products and markets, drives a global economy? What else but insatiable curiosity gives rise to the oversupply of knowledge-technological objects and the new knowledge with which science and technology create the preconditions for further innovation activity, even if the path to it is long and never direct?

But answers like these ignore the far-reaching changes that the production of knowledge and the science system are exposed to today. Such answers stop at the surface. They neglect the political economy of science,[94] which, after the end of the cold war, was placed on new foundations that have a far-reaching effect on the organization and funding of research. The (relative) withdrawal of the state and with it of public funding for research has led to a wave of privatization that carries issues of intellectual property rights into every laboratory. Every university must face and must orient its management toward the question of its profile and its strategic goals. Industry's support for university research has reached a strategic level. It is oriented toward, among other things, how well the universities can place their research internationally and regionally and how they can assert themselves in global competition for the best young scientists and engineers. Research policy on the levels of the European Union and the nation-state is pressing for better and faster implementation of scientific results in palpable competitive advantages on the market. It is accepted that basic research works in the longer term, but the expectation is clear that it, too, should contribute to use, to the implementation of knowledge in innovations.* Political economy has caught up with science and the production of knowledge for good.

But things are not as simple as they may appear. Analogies can be drawn between (1) Schumpeter's distinction between inventions or the idea of the production of something new and innovation as the successful implementation and utilization of ideas and inventions in society and (2) evolutionary biology (for example, between the physical causes of genetic and phenotypical variability in organisms and the factors that lead to the fixation of a preferred variant within a population).[95] Coarsely simplified, invention can be characterized as the starting point, and innovation as the result and success. But neither in the history of evolution nor in human societies does the rise of a new quality or the selective production of a new combination permit predictions about the evolutionary, ecological, or cultural effect. Invention is a weak predictor of success.

So attention is directed toward everything that creates connections, gives rise to networks, or promotes interactivity. Producers and users belong to the same community when it is a question of promoting exchange and feedback that contribute

* As Robert K. Merton remarked as early as 1957, scientists were long convinced that the social effects of their research *have* to be useful in the long term. This credo had the function of legitimizing research activity but in fact is easy to disprove. It shows the mixture of truth and social utility that Merton termed the "non-logical margins of science." See Robert K. Merton, *Social Theory and Social Structure* (New York: Free Press, 1957).

to the improvement of products or inventing new applications. Communications and information technologies connect people and technological things technologically, socially, economically, and culturally. Their future lies in inventing and implementing new forms of interactivity, permitting an individual to communicate with many people—and many people to communicate with many people. Everywhere, the users, whether they already see themselves as such or not, are to be involved. In biology, the old dichotomy between nature and environment is being redefined as an evolutionary process of niche construction. For example, if earthworms consume a certain number of tons of earth per hectare and year, the amount of carbon and nitrogen in the soil increases, which benefits the earthworms and other organisms. Invention and innovation are connected by means of a complex feedback system that eliminates the rigid boundaries between the environment and the organism, between development and selection. Organisms actively contribute to shaping their environment; they influence their own selection regime, just as users are actively involved and themselves become producers of new use functions.

How did it come about that we are now all "empowered" to build, like the earthworms, our own economic and cultural niche, to found our own companies, and to become the entrepreneurs of our own occupational biography and the directors of the numerous stagings of our selves, all in the knowledge that the probability of failure is great? Why are we willing to take part in television games and other media-conveyed interactions that simulate forms of communication that we know are not authentic—even if they are real in the sense of another kind of conversation, encounter, or creation of interpersonal relationships? Where is the resistance against the unfamiliar, the other, when we willingly accept what medicine offers us to provide our defective or aging bodies with replacement parts that are artificial or taken from other organisms? We see through the illusions that we thereby submit to. Nevertheless, we cling to them to continue building the niches in full knowledge that we can help shape them and are at the same time sobered in the knowledge of the limits to our self-determination and our possibilities of control.

We have remained modern—and will have to remain modern in the future, as well. Part of this is the deep-seated discontent and the odd ambivalence that consists in recognizing what we have created on our own and at the same time experiencing it as something external, alien, constraining, or limiting. Constructing a niche means accepting the unforeseeable interaction with what is arising and how it arises. If it is true that the brain construes a reality that seems stable to us, then perhaps the institutions that we have created also aim to create a social reality that seems more or less stable to us. The ambivalence that is a characteristic of modernity is a response to the tension underlying every controversy, every dialogue, every painstaking cogitation about the relationship between society and science. It is the tension between the demand for autonomy, self-determination, and human freedom and the inescapable fact of limitation, loss of control, and hegemony. On the one side is our knowledge about ourselves, our experience, our demands, and our wishes; on the other side is our knowledge of what is not ourselves, of the others, and of the world as it exists.

This contradiction and the friction resulting from it stood in the center of the creative work of modern art. The sensibility of the modern subject, which so often manifests itself as discontent with modernity and as a form of rebellion against the material and institutional achievements that in turn have created the preconditions for the existence of the modern subject, has become an integral component of the project of modern art. From Baudelaire through Nietzsche to the representatives of postmodernism, a thread of discontent, inner conflict, and ambivalence runs counter to the technological optimism of their respective times. In painting, music, literature, and avant-garde philosophy, being modern is experienced as a regrettable or terrible fate and treated accordingly. According to Robert Pippin,[96] many of the questions experienced as so problematical have less to do with traditional aesthetics than with the wrestling to understand the changes in space and time: how can historical time be lived at all, and with what means is artistic creation able to express itself in temporality? The question addresses the spatial anchoring of the self and of art at a moment that itself cannot be localized because the radical quality of the break carried out by modernity seems to prevent localization.

From this and many other artistic and literary testimonies speaks the voice of the despairing individual who is in conflict with herself and whose failure to achieve the ideals and aspirations she has set for herself can turn into a kind of contempt and self-hatred. The contrast to the free, rational, independent, self-determined, and self-determining subject could not be greater. It was left up to art and literature to find out which of the demands made could be successfully met and which were condemned to failure. The demands were primarily those that the individual placed on herself but also on the others and the society, and the artistic and literary experiments served to explore the complex social and existential dependences.

The modern bourgeois self-hatred is a phenomenon that arose historically in a niche and that built itself a niche. The impetus arising from the discrepancy between dissatisfaction with what one has experienced and the aspirations one has set for oneself and that were experienced as very risky to implement can take many different forms, including political forms, as the social and political movements of the nineteenth and twentieth centuries demonstrated. But as Pippin elucidates in his work *Modernism as a Philosophical Problem: On the Dissatisfaction of European High Culture*, these aspirations come from deeper layers. For modernity, it is the problem of freedom. Applied to the individual, this means the freedom to live in accordance with one's own ideas and to identify with one's own deeds and actions. It is the freedom to accept one's deeds.[97]

The radical breaks and changes that modernity brought and brings with it point in different, even seemingly opposite directions. On the one hand, an ever-tightening network of mutual dependences arises,[98] among them the growing dependence on technology and on a culture molded by technology that sets standards of behavior, perception, and thinking. On the other hand, modernity brought unimagined freedoms that accompanied the breakup of usually close social communities, a political and social emancipation that, with the aid of technology and science, has expanded the radius of action for spatial and intellectual mobility and that, under the vague concept of self-realization, integrated the body and social identity in the modern imagination of shaping oneself. The late nineteenth and early twentieth centuries are full of impressive literary, historical, and sociological testimony that took this process of compact interpenetration—the social compression as well as the condensation of time and space—as the object of its observation and analysis. The social pressure of the small community, whether the village or the narrow social circles of the upper classes, gave way to another kind of societal pressure to conform, one guided by economic success. It gave rise to the fears related to one's own authenticity, to the "true" self that is at the center of many literary works.

The artistic and literary delving into themes like human and other organisms' genetics, the erosion of boundaries that this leads to, and the struggle to find a language allowing the definition of human characteristics and new identities continues unabated.[99] The free spaces that seem likely today promise unimagined possibilities of a neurochemically steered enhancement (and some would say, invention) of the self and of a changed (because more intimate) way of dealing with technology that is integrated in the body, is mounted or worn on the body, and becomes the manmade environment. But however seductive this interactivity-created intimacy might be, it presupposes the voluntary abandonment of knowledge of and control over the self. The view from inside must learn to interact with the view from outside and if possible to converge in a way that can societally stabilize the new knowledge-technological reality. The discussion about freedom of the will, for example, which was conducted on a sophisticated level in the features sections of newspapers, tried to break up part of the hegemony of the neurosciences by considering other forms of knowledge and the claim to autonomy. This can be achieved only if the ambivalence underlying the tension is permitted to become visible and is recognized as legitimate. The artistic and literary elites of the nineteenth and twentieth centuries mostly concluded that the concept of freedom and the premonition of the realization of a potential inherent in human beings to work together and to live together in peace are nothing but illusions. The findings are not yet in on the discourses and controversies that are being carried out today, but the verdict could be similar here as well, this time coming from the natural scientific viewpoint.

But modernity also consists in permitting varying viewpoints. The natural sciences must be prepared to accept part of the I-perspective; vice versa, the self with its personal experience must accept the way it is perceived from the perspective of natural science. But making judgments and decisions cannot be reserved for the experts. Modern societies have brought forth various forms of modern agendas, and modernity is not homogeneous. On the contrary, it consists of "multiple modernities."[100] The intensified discussion between society and science makes many of the beaten tracks of dichotomies and differentiations (for example, between "rational" and "irrational") outmoded. What is needed instead is more differentiation, more reflectivity, and the ability to see things from the standpoint of the other side. There is also more than one way of being scientific.

Another piece of evidence for multiple modernities comes from the inclusion of the public as part of the expanded understanding of democracy. The discontent with today's modernity has shifted from literary testimony to the public's demands to be permitted to take part in decision-making processes that involve complex scientific-technological content and in setting research priorities and the future orientation of the production of knowledge. Register, site, and protagonists have thus changed. At the same time, bourgeois self-hatred has vanished with the bourgeois individual. In the same way, the tension between experience and expectation is defined by different parameters. Expectations that many people used to project beyond their own lifetimes have turned into things taken for granted. The extension of life expectancies creates new problems for aging societies. The historical experience of acceleration* and other effects of science and technology on life and work have set the spiral of wishes moving upward and at the same time created an instantaneous, if only brief, wish-fulfillment. The time periods within which expectations can solidify are shortening. The feeling of living today for tomorrow is spreading, and uncertainty is increasing. What remains is the deep-seated ambivalence, the characteristic of modernity.

* Cesare Marchetti, who for thirty years has investigated the regularities of the rise and satiation of competing systems, asserts that means of transportation like airplanes do not primarily aim to save time but that the majority of people, "from the Zulus to the American upper class," spend an average of sixty-five minutes a day changing their location. Growing income may lead to the use of faster means of transportation, but primarily this merely extends the radius of the territory to be put behind one. Using the airplane does not so much save time as it permits the "control" of a larger area. Cesare Marchetti, *Logos, il Creatore di Imperi* (IIASA, Interim Report IR-04.043, September 2004).

What about our expectations today? In Europe, the major project of modernity, with its bundle of promises that were coupled with diffusely articulated but actually held expectations, passed its zenith in the 1970s. What followed were more negatively defined expectations that aimed to secure what had already been achieved—the avoidance or reduction of risks of various kinds, as articulated primarily by the protest movements that were preceded or driven by the risk society. As guidelines for future action, they can be reduced to a few principles. They are defined less by their content than by processes borne by the insight that the constantly changing content cannot be grasped—the principle of sustainability in our dealings with the vanishing resources of our natural environment and the precautionary principle in our dealings with uncertainty. But it is difficult to declare the instantaneous gratifications of the consumer society to be expectations, even if the indicators of market research label them that way. The horizon of expectations was leveled in the face of an oversupply of renewals in the search for latent desires and needs.

Of course, basic expectations remain—the concern for health, security, community and solidarity, and the fulfillment of the individual's own inherent potential. The greater the changes, the stronger the wish for something change-resistant to hold onto. Where expectations and experience converge too closely and the open horizon of the future either shrinks to a tiny gap (allowing in only what passes the risk-precaution test) or suffocates on the oversupply of products that all resemble each other: this is where we observe the increasing attraction of a past that never existed but that is all the easier to imagine.

But by themselves, the knowledge-technological visions do not contain an adequate image of human beings, much less an image that gives confidence.

The speed with which innovations arrive on the market leaves little time for enthusiastically greeting them. The movements of the first little robot probes on Mars were compared with those of a toy, and even the planet's stony surface seemed familiar since it resembled that of the Atacama Desert in Chile. The spectacular and wonderful images of Saturn's rings that were broadcast to Earth aroused fervent admiration, but they soon landed in the archive of visual memory in which a scientific, multimedia virtual reality must also compete for public attention. The U.S. president's announcement that he wanted to have the American nation conquer space and to this purpose planned a middle-term moon expedition left even the experts cold (though for other reasons). A British commentator complained that such a mission belonged to the "past century." The first moon landing stood for the success of Western technology over that of the former Soviet Union, a symbol of the free world that could be transformed into consumer goods and corresponded to the optimism of the 1970s, but little of this is left at the beginning of the twenty-first century.[101]

Science and technology cross the boundary between the present and the future with a certain ease and thereby move the future closer to the present. Nonetheless, the future seems fragile. The loss of temporal distance blurs the difference between what is technologically possible and what is already present in the laboratory, between imagination and reality, which is often a virtual reality. Having lost all utopias, the future presents itself as a sketch of technological visions that block out the social knowledge that is needed to live in a scientific-technological world—and to feel well in it.

It is no coincidence that the discrepancies among our claims to our own body and our power over it, the definition of the self, and the possibilities of intervention and alteration that result from the achievements of molecular biology, genetics, reproductive medicine, and other areas of the life sciences acutely reveal the contradiction between autonomy and dependence. This contradiction is promoted by two processes coming from different directions and meeting in the field of tension between science and the public sphere. The first receives its impulses from science itself or, more precisely, from the transformations that the science system is currently going through as a result of its increasing interlocking with society. Science is becoming societally contextualized,* which means that prioritizations within research—the relationships to use and application that, even in basic research, are at least potentially felt out and created, even as far as the specific determination of which living organisms under what conditions, what processes for creating new phenomena, and what hybrid constructions are subject to property rights—are coproduced by societal, economic, political, cultural, and ethical framework conditions.

Newly gained knowledge and new research instruments usually lead to new questions and problems that hadn't been

* As was already the case in Galileo's time. A changed context brings different societal values, preferences, and current framework conditions.

posed before. The capacity to sequence genes raised questions of their patentability and the associated property rights. The possibilities arising from the acquisition of embryonic stem cells trigger heated moral debates. Which procedures resemble legal therapeutic interventions that are already used and which procedures place one kind of parenthood, social parenthood, above the other genetic kind? What possibilities are there to shape human reproduction democratically, assuming that the safety of the procedure can be guaranteed?[102] Or what new, efficient technologies make it possible to analyze and screen large amounts of genetic data, large-scale projects that involve whole segments of the population for years and make it urgently necessary to answer questions of protecting the private sphere and the public interest or of the possibilities of interfering in existing rights?*

But progress within science is not the only thing that drives the process of contextualization forward. If research is today regarded as the driving force of economic competitiveness, then corresponding research investments must be made. When support from public coffers stagnates, competition for private research investment ensues. The latter often concentrates on the economically promising areas in the high-tech sector and in the life sciences. Forms of so-called public-private partnerships are fostered, and new forms of research organization develop. With increased private investment, the issue of

* For example, when volunteers in a control group in longitudinal studies are expected to eschew in advance certain patients' rights in the interest of a randomized clinical experiment.

intellectual property rights to research procedures and their results (for example, artificially produced life forms and organisms, gene sequences, and various sets of data) takes on acute significance. Like all property rights, intellectual property rights regulate relationships between people and not between people and things. They ensure that ideas, procedures, or organisms placed under limited-term protection can be used only after paying appropriate fees. The protected objects remain in the public sphere, but access to them and the way they can be used are subject to certain regulations and economic limitations. It is interesting to see that, under the pressure of increasing privatization, researchers are beginning to see themselves as the proprietors of their data and their research results.[103] They no longer want to be "knowledge workers" but "knowledge owners."

But the change in the foundations regulating the exchange of information and knowledge in the academic world, in which mutual trust is in part being replaced by contracts, is just one side of the coin. Science is also coming under pressure from another side. This arises in the name of democratization, which, with its demands for a voice for the public in complex decision-making processes that crucially affect that public, does not stop at the institution of science. Science's (relative) freedom of action and its culture of autonomy are coming under the pressure of accountability to both the organizations that promote research and also the public. It is not enough for science to maintain relationships to the state and the market; it should also give civil society and its needs a place in the research agenda and in its mediation.

Privatization and democratization are interconnected in that both stand for the success so far of the highly industrialized

Western democracies. Property rights and privatization stand for a regime that is regarded as an integral component of the Western industrial states' success in economic growth. It is expected of science and technology, as the driving forces of this growth, that the principles underlying the increase of economic productivity can be profitably transposed to the production of new knowledge. The efficiency of markets, competition, institutions, and intellectual property rights is expected to develop its productivity-increasing effect also where the new knowledge, epistemic things, abstract objects, and symbolic technologies (which are ultimately the innovative potential) are produced. Equally, democratization of scientific expertise is the expansion of principles of governance that, up to now, have been serviceable to the liberal Western democracies. Their citizens have achieved a historically unique level of education. They do not permit themselves to be overawed by scientific-technological achievements, however great they may be, nor are they willing to accept the opinions of scientific experts uncritically. By insisting that it be accorded adequate political representation in important decision-making processes, the public[104] answers with a loud and politically articulate voice, just as investors and markets are in dialogue with science when they call for greater efficiency and productivity gains in exchange for private research investments. Both processes put pressure on science and change its nature as the public good it has been regarded, as least historically.*

Unlike in the nineteenth and early twentieth centuries, the struggle for the autonomy of the free individual no longer

* In this form, "nature" is as nonexistent as "society."

takes place primarily in art, literature, or philosophy. The private fears and defense strategies of a small cultural and social elite, as significant as its achievements are, have been supplanted by the everyday fears of the great majority in the Western liberal democracies with their scientific-technological civilization. Now, issues like protecting the private sphere and determining whether and to what degree property rights are possible in relation to the human body and its genetic substance are being thrashed out on the stage of everyday life. The discrepancy between, on the one hand, the claim to and the determination of one's own body and the definition of the self and, on the other hand, the possibilities of intervention and alteration that result from the achievements of molecular biology, genetics, reproductive medicine, and other areas of the life sciences starkly reveals the contradiction between autonomy and dependence. Heated controversies have broken out over the definition of human life and the moment when it begins and ends. The discussion of risk focuses on the distinction between risks that the individual takes knowingly and voluntarily and risks that appear involuntarily and collectively. The conflict is over citizens' ability to judge and decide in comparison with the evaluation of the judgment of experts. At stake is the constitutionally anchored freedom of inquiry and its possible limitation in the interest of other democratic rights. Compromises must be found for value conflicts in which religious and ethical standpoints are irreconcilable. The freedom that the one side demands is rejected as the irresponsible limitation of the freedom of the other side. "Society" has taken the place that was once reserved for an artistic elite. The discontent with modernity is part of modernity, as is democracy, which has become a medium of expressing and negotiating this discontent.

The controversies are not about science fiction scenarios but about decision-making situations and dilemmas that have begun to play an uncomfortable and unsettling role in everyday life. They call for a normative foundation, a basic consensus (which is hard to achieve), as well as legal and political regulations and their implementation. What the artistic and literary sensibility of modernity eloquently expressed had reached everyone by the beginning of the twenty-first century. Nonetheless, the heroic confrontation with the impossibility of living one's own ideals and aspirations and of striving for an evaporating authenticity continued in what many experienced as the tensions of dealing with uncertainties and of asserting the self in practical life despite existing dependencies.

The question of whether it is possible to acquire too much knowledge—more than is good for people—seems a thing of the past. Roger Shattuck meticulously narrated the history of forbidden knowledge, which is simultaneously a ban on knowing,[105] and Hans Blumenberg set a lasting monument to it with his history of modern times as the history of newness.[106] But the triumphant march of a secular science and its monopoly on explanations of the natural world stifles the questions of the natural limits of knowledge and of being permitted to know. The ambivalence also manifests itself in the concealed appeal to nature as the umpire over human *libido sciendi*, which violates the old religious taboo against wanting to know more than God in his omniscience permits.

But nature has no moral principles and does not know the future tense. Evolution, to which the expansion of human knowledge is also owed, knows neither set limitations nor a final destination. In the current debates on stem-cell research, genetically modified organisms, and the risks from nanoparticles, one thing constantly appearing is a recoiling from human knowledge and its consequences. This is not only an expression of the public's distrust of science and of political institutions, as the results of opinion polls might suggest. The dithering and shrinking back from the consequences of curiosity also testify to the wish for reassurance in the face of a seeming excess of uncertainty. The ambivalence of modernity appears again. This time it is rooted in the impossibility of making final statements about what segment of the world our knowledge enables us to understand and what being human means in a world that we ourselves constantly change. It is inherent in the preliminary quality, the temporal context of all knowledge and in the difficulty in accepting this preliminary quality.*

The ambiguity also applies to the insatiability of the wishes and the desire inherent in the promises of modernity. Henri Lefevbre writes that the growth and appropriation of desire and

* This explains the struggle for perception, for a conceptual understanding of the self and of identity, and for the possibilities of intervention and change in this self's relationships as part of a cultural, humanmade evolution. That this ambivalence and its forms of expression are historically and culturally mutable has been shown by the anthropologist Pamela Asquith's observations of the burials that Japanese primatologists have carried out for the monkeys killed in their research work. Pamela Asquith, *The Monkey Memorial Service of Japanese Primatologists*, D. Phil. thesis, Oxford University, 1981.

wishes are also parts of modernity.[107] The freedom to want ineluctably raises questions of morality—of what the nature of our positive, meaning-creating dependence on others is and what we owe them in the light of this bond, as Robert Pippin aptly put it.[108] The yearning or striving for freedom and for leading a life that corresponds to one's own ideas necessarily has to deal with the wishes that others have and the ways that their striving with freedom is compatible with one's own. In whatever this freedom consists—either material independence and money or power and thus the ability to evade the will of others—it cannot be achieved solely for oneself.

In *Tomorrow's People,* a book written for a broad readership, Susan Greenfield shows a future in which people no longer wish for anything.[109] The renowned British neuroscientist asks about the effects that interactive biotechnologies currently being worked on will have on human thinking and feeling. She thereby covers a broad spectrum—from the lifestyle of the future through how we will perceive reality, how we will think about the body in the face of progress in robotics, how we will spend our time, how we will work, how we will love, what family structure is probable, and what and how we will learn (including how to live with terrorism). But as scientifically and technologically plausible as the depicted future may be, it appears emptied of meaning and unworthy of striving for. The author considers it possible that a change in neuronal consciousness will bring a loss of human individuality, which will dissolve in a kind of hedonistic homogeneity. Since all wishes will be fulfilled, wishes will disappear, and the private sphere will make way for an all-embracing, self-organized collectivity.

But the author falls into the trap of her own naturalistic description, which is not societally situated despite the selected examples. In this world of tomorrow, there are no politics and no economics. It remains unclear who works for whom. Culturally, hedonism dominates. And yet it is very clear that *Tomorrow's People* greatly disturb her. She tries to rescue the situation by making out an unexpected succor near the end of the book. This is to come from the "great majority"—the millions of people in this world who are still hungry, who cannot afford and do not want the technological luxury of the rich north, and who (she speculates) may prevent us from destroying human individuality. The people of tomorrow whom Greenfield presents are pitiable creatures. They have left modernity behind them and have lost the capacity to imagine themselves in others' situation and to understand others. With that, human diversity ends. And since these people of tomorrow are no longer able to understand each other, they forget how to deceive, defraud, and pretend to others. They lose the feeling for their own identity because they can no longer recognize what is "genuine" or "authentic" about themselves or whether they are ultimately controlled or programmed by institutions; they are thus other-directed. They lose the feeling for difference and for the ambivalence of modernity.

In her vision, thinking about the future will have become empty at the same time as it is overfull in comparison with the exuberant imagination with which looking into the future enlivened the past. The imaginary space loses what once gave it its attractiveness and vitality—that it was compensation for the present and a projection screen for hopes and fears that were connected with the granting of a postponement of all the unfinished agendas of the present. But this emptiness of the future, this poverty of societally relevant imaginative power, confronts a flood of information about technological and scientific innovations that have not yet found a place in the society of tomorrow and have not yet been accepted by the people of tomorrow. The language and the images that the innovations use are not necessarily created by the natural sciences and technology. They bear the latter's trademark and claim that only they represent reality. But in precisely this point, they neglect the ambivalence of modernity. They ignore cultural phenomena whose assumptions are rooted in their claim to be part of reality. They deny the need for a change in the culture of consciousness,[110] a change in the image of human beings, and a scientific-technological culture that does justice to contradictory viewpoints and knows how to relate them reflectively to each other.

The future and the people of tomorrow are still primarily conceived in utopian and dystopian images whereby utopianism makes use of the genre of scientific-technological visions and their unconditional enthusiasm while dystopianism prefers the literary or artistic narrative form and posits that things are headed for catastrophe. But both the scientific-technological visions and their complement, the dystopian images of the future, attempt to suppress the ambivalence of modernity. This ambivalence teaches us that the people of tomorrow will no longer be the people we know today. Nor will they be cyborgs and androids, the hybrid figures of science fiction, who fascinate us because we do not know the ways in which they resemble and differ from people like us. To understand them, we must put ourselves in their place and estimate the possible effects of our actions on them. In this way, we make another of the many attempts in history that have been made to find a foundation for our own behavior—a foundation that asks what the nature of our positive, meaning-creating dependence on others is and what we owe them in the light of this bond. Ultimately, this is the only way we can be self-determining and know who we, the people of today, are. If we want to conceive the future outside of the categories of utopia and dystopia, we have to start out from the people of today.

Technological systems require a degree of compatibility in their standards and components that the social systems cannot have because they must remain open. We expect that technological systems must be foreseeably reliable and secure. Only then are the selected technological solutions stable enough to solve the problems posed to them. By contrast, social systems—and societies—constitute themselves from their members' knowledge of each other. They are not subject to any process of closing and must remain in continuous openness. We know what the world's top scientific laboratories are working on today, and yet at best this allows us to derive scientific-technological visions that fit within a system that has been made to be consistent within itself. These visions can say next to nothing about forms of social organization, mutual relations among people, and emotional energies that the people of tomorrow will invest in ideas for or against each other or in things and institutions whose continuity they believe in. This lack of social knowledge makes these technological visions blind, even if they are able to gauge a limited number of "impacts"—of foreseeable effects whose corresponding consideration ought to be self-evidently a component of the process of generating technology. For as John Maynard Keynes remarked, the unavoidable never happens, while the unexpected always occurs.

This knowledge that the people of today have from and about each other includes information and knowledge of existing mutual dependence. It is tied to collectively acknowledged norms that are subject to a historical process of change. These norms presuppose the ability to judge and the knowledge of when and how they are to be applied. Collective attribution of meaning grows out of a culture shared with others, which in turn presupposes normative obligations and collective experiences. Innovation as the collective bet on an uncertain future therefore cannot content itself with technological visions. However proper it may be to work from the great natural scientific discoveries of the twentieth century, as represented by atoms, computers, and genes, and to conclude from them that the quantum revolution, the computer revolution, and the biomolecular revolution will shape all further scientific-technological achievements through to the year 2050, this in no way implies that we can see in this the foundation for the beginning of an imminent planetary civilization.[111] The vision remains one-eyed even if it professes to look far ahead. Predicting the broad lines of development (for example, the convergences already taking shape between bio-, information, and nanotechnologies) does not yet mean being able to predict the outlines of the society in which the people of tomorrow, who are just now constructing their next niche for themselves, will live.

That is why innovation cannot be oriented toward a specific goal. It is a process in which the space of possibilities is opening up and opportunities that usually arise unexpectedly should be used. As a process, innovation is never temporally or spatially finished. It is something preliminary whose dynamic pushes forward but knows no end point or arrival. It is thus more self-sufficient and at the same time more pragmatic than the technological visions that its pragmatism includes. Breakthroughs occur; they cannot be planned. Failures are part of the learning process, and the possibility of failure is always present. And so is the possibility of tipping over, standstill, or stagnation. Society can refuse, and resistance can grow strong. Will reality in the end imitate theory, as happened in the financial markets? The answer will depend on whether we are willing to remain modern in the future—for a future that, as the construction of a special kind of niche, requires ambivalence as a cultural resource.

Notes

Chapter 1

1. Appadurai, Arjun. 2004. "The Capacity to Aspire: Culture and the Terms of Recognition." In: Vijayendra, Rao and Walton, Michael (eds.), *Culture and Public Action*. Stanford: Stanford University Press, pp. 59–84.

2. Pippin, Robert. 2005. "Introduction: 'Bourgeois Philosophy' and the Problem of the Subject." In: *The Persistence of Subjectivity: On the Kantian Aftermath*. Cambridge: Cambridge University Press, pp. 14–15.

3. Esposito, Elena. 2004. *Die Verbindlichkeit des Vorübergehenden: Paradoxien der Mode*. Frankfurt am Main: Suhrkamp.

4. Luhmann, Niklas. 1986. *Ökologische Kommunikation. Kann die moderne Gesellschaft sich auf ökologische Gefährdungen einstellen?* Opladen: Westdeutscher Verlag, p. 320.

5. Maynard Smith, John and Szathmary, Eörs. 1999. *The Origins of Life: From the Birth of Life to the Origin of Language*. Oxford: Oxford University Press.

6. Toulmin, Stephen. 1991, 1994. *Kosmopolis. Die unerkannten Aufgaben der Moderne*. Frankfurt am Main: Suhrkamp.

7. Luhmann, op. cit., p. 161.

8. Jacob, François. 1983. *Das Spiel der Möglichkeiten. Von der offenen Geschichte des Lebens*. Munich: Piper Verlag, p. 94.

9. McSherry, Corynne. 2001. *Who Owns Academic Work? Battling for Control of Intellectual Property*. Cambridge, Mass.: Harvard University Press.

10. Nowotny, Helga. 2005. "The Changing Nature of Public Science." In: Nowotny, Helga et al., *The Public Nature of Science under Assault: Politics, Markets, Science and the Law*. New York: Springer.

11. Leroi-Gourhan, André. 1980. *Die Hand und das Wort. Die Evolution der Sprache, Technik und Kunst*. Frankfurt am Main: Suhrkamp, p. 270.

12. Donald, Merlin. 2001. *A Mind So Rare: The Evolution of Human Consciousness*. New York: Norton.

13. Epstein, Steven, oral communication with the author.

14. Greenfield, Susan. 2003. *Tomorrow's People: How 21st-Century Technology Is Changing the Way We Think and Feel*. London: Allen Lane, Penguin Books.

15. Ezrahi, Yaron. 2003. "Science and the Postmodern Shift in Contemporary Democracies." In: Joerges, B. and Nowotny, H. (eds.), *Social Studies of Science and Technology: Looking Back, Ahead*. Dordrecht: Kluwer Academic Publisher, pp. 63–75; Ezrahi, Yaron. 1990. *The Descent of Icarus, Science and Contemporary Democracy*. Cambridge, Mass.: Harvard University Press.

16. Markl, Hubert. 2001. "Research Doesn't Denigrate Humanity: Being Human Is More Than Simply Having the Right Molecular Composition." In: *Nature*, vol. 412, pp. 479–480.

17. Daston, Lorraine. 1992. "Objectivity and the Escape from Perspective." In: *Social Studies of Science*, vol. 22, pp. 597–618.

18. Schaffer, Simon. 1997. "What Is Science?" In: Krieg, John and Pestre, Dominique (eds.), *Science in the Twentieth Century*. Amsterdam: Harwood, pp. 27–42.

19. Browne, Janet. 2003. *Charles Darwin: The Power of Place*. Princeton: Princeton University Press, p. 56.

20. Golinski, Jan. 1998. *Making Natural Knowledge: Constructivism and the History of Science*. Cambridge: Cambridge University Press, p. 191.

21. Beer, Gillian. 1994. "Against the Grain: Thinking across Nature." In: Nowotny, Helga (ed.), *Innovation: Kreativität in Kunst und Wissenschaft. Ergebnisse des Initiativ-Workshops*. Vienna: IFK Materialien, pp. 15–19.

22. Beer, op. cit., p. 18.

Chapter 2

23. Stengers, Isabelle. 2002. *Penser avec Whitehead. Une libre et sauvage creation des concepts*. Paris: Editions du Seuil.

24. Stengers, ibid., p. 18.

25. Latour, Bruno. 1993. *We Have Never Been Modern*. Cambridge, Mass.: Harvard University Press, pp. 57–58. In German: 1998. *Wir sind nie modern gewesen. Versuch einer symmetrischen Anthropolgie*. Frankfurt am Main: Fischer Verlag.

26. Haldane, J.B.S. 1924. *Daedalus, or Science and the Future: A Paper Read to the Heretics*. Cambridge, February 4, 1923.

27. Haldane, ibid, p. 46.

28. Russell, Bertrand. 1924. *Icarus: or, the Future of Science*. London: Kegan Paul, Trench, Trubner, p. 58.

29. Ezrahi, Yaron. 1995. "Haldane between Daedalus and Icarus." In: Krishna, R. Dronamraju (ed.). *Haldane's Daedalus Revisited*. Oxford: Oxford University Press, pp. 64–78.

30. Pavitt, Keith. 2004. "Changing Patterns of Usefulness of University Research. Opportunities and Dangers." In: Grandin, Karl, Wormbs, Nina, and Wildhalm, Sven (eds.), *The Science-Industry Nexus: History, Policy, Implications*. Sagamore Beach: Science History Publications/USA and the Nobel Foundation, pp. 119–131.

31. Piccinini, Patricia. 2003. *We Are Family*. First published by the Australian Council. Australian pavilion, Fiftieth Venice Biennial.

32. Blumenberg, Hans. 1966, 1996. *Die Legitimität der Neuzeit*. Frankfurt am Main: Suhrkamp.

33. Gardener, Howard. 1982. *Art, Mind, and Brain: A Cognitive Approach to Creativity*. New York: Basic Books.

34. Schmidgen, Henning, Geimer, Peter, and Dierig, Sven (eds.). 2004. *Kultur im Experiment*. Berlin: Kadmos.

35. Rheinberger, Hans-Jörg. 2001. *Experimentalsysteme und epistemische Dinge. Eine Geschichte der Proteinsynthese im Reagenzglas*. Göttingen: Wallstein Verlag.

36. Rheinberger, ibid., p. 249.

37. Julius H. Comroe, Jr. and Robert D. Dripps. 1975. "Ben Franklin and Open Heart Surgery." *Circulation Research*, vol. 35, pp. 661–669, pp. 105–111 (quoted after Stokes).

38. Comroe and Dripps, ibid., p. 106.

39. Stokes, Donald. 1997. *Pasteur's Quadrant: Basic Science and Technological Innovation*. Washington, D.C.: Brookings Institution Press.

40. Holton, Gerald. 1986. *The Advancement of Science and Its Burdens*. Cambridge, Mass.: Harvard University Press; Holton, Gerald, 1998. "Jefferson, Science, and National Destiny." In: Berlowitz, L. et al., *America in Theory*. Oxford: Oxford University Press.

41. Nolte, Paul. 2004. "First Contact. Die Tagebücher der Lewis & Clark-Expedition und die bescheidenen Anfänge des amerikanischen Expansionsdranges." In: *Literaturen*, 7/8, pp. 32–36.

42. Stokes, op. cit.

43. Perkins, D. N. 1992. "The Topography of Invention." In: Weber, R. J. and Perkins, D. N. (eds.), *Inventive Minds: Creativity in Technology*. New York: Oxford University Press, pp. 238–250; Perkins, D. N. 1994. "Creativity beyond the Darwinian Paradigm." In: Boden, M. (ed.), *Dimensions of Creativity*, Cambridge, Mass.: MIT Press, pp. 119–142.

44. Perkins, David. 2000. "The Evolution of Adaptive Form." In: Ziman, John (ed.), *Technological Innovation as an Evolutionary Process*. Cambridge, Mass.: Cambridge University Press, pp. 159–173.

45. Gras, Alain. 2003. *Fragilité de la puissance. Se libérer de l'emprise technologique* Paris: Fayard.

46. Hughes, Thomas P. 1987. "The Evolution of Large Technological Systems." In: Bijker, W., Hughes, T. P., and Pinch, T. (eds.), *The Social Construction of Technological System*. Cambridge, Mass.: MIT Press, pp. 281–292.

47. Nowotny, Helga. 1999. *Es ist so. Es könnte auch anders sein. Über das veränderte Verhältnis von Wissenschaft und Gesellschaft*. Frankfurt am Main: Suhrkamp.

48. Scott, James. 1998. *Seeing Like a State: How Certain Schemes to Improve the Human Condition Have Failed*. New Haven: Yale University Press.

49. Bernal, John Desmond. 1939, 1986. *Die soziale Funktion der Wissenschaft*. Cologne: Pahl-Rugenstein.

50. Nowotny, Helga. 2005. "The Changing Nature of Public Science." In: Nowotny, Helga et al., *The Public Nature of Science under Assault: Politics, Markets, Science and the Law*. New York: Springer.

51. Appadurai, Arjun. 2001. "Grassroots Globalization and the Research Imagination." In: Appadurai, Arjun (ed.), *Globalization*. Durham, N.C.: Duke University Press, pp. 1–21.

52. Nowotny, Helga, Scott, Peter, and Gibbons, Michael. 2001. *Re-Thinking Science: Knowledge and the Public in an Age of Uncertainty*. Cambridge: Polity Press. In German: 2004. *Wissenschaft neu denken. Wissen und Öffentlichkeit in einem Zeitalter der Ungewissheit*. Weilerswist: Velbrück Wissenschaft.

53. Bijker, Wiebe. 2005. "The Vulnerability of Technological Culture." In: Nowotny, Helga (ed.), *Cultures of Technology and the Quest for Innovation*. New York: Berghahn, pp. 52–69.

54. Durkheim, Émile. 1915. *Elementary Forms of Religious Life*. London: Allen & Unwin, p. 56.

55. Nowotny, Scott, and Gibbons, op. cit., 2001, 2004.

56. Nowotny, Scott, and Gibbons, op. cit., 2004 (German version).

57. Bloor, David. 1976, 1991. *Knowledge and Social Imagery*. Chicago: Chicago University Press.

58. I owe this idea, like many others, to Marilyn Strathern. See Strathern, Marilyn. 2004. *Social Property: An Interdisciplinary Experiment*. In: *PoLAR*, vol. 27, no. 1, pp. 23–50.

59. Lloyd, Geoffrey and Sivin, Nathan. 2002. *The Way and the Word: Science and Medicine in Early China and Greece*. New Haven: Yale University Press.

60. Vogel, Hans Ulrich. 2005. "The Mining Industry in Traditional China: Intra- and Intercultural Comparisons." In: Nowotny, Helga (ed.), *The Quest for Innovation and Cultures of Technology*. New York: Berghahn.

61. Julien, François. 1996. *Traité de l'efficacité*. Paris: Grasset & Fasquelle.

Chapter 3

62. Schatz, Gottfried. 2004. "Was hemmt die Innovation in der Schweiz?" *NZZ* Sept. 18–19, p. 45.

63. Fagerberg, J., Nelson, R., and Mowery, D. (eds.), *The Oxford Handbook of Innovation*. Oxford: Oxford University Press.

64. Kocka, Jürgen. 2001. *Gebhart. Handbuch der Deutschen Geschichte. Das lange 19. Jahrhundert*, vol. 13, pp. 38–39.

65. Mithen, Steven. 2003. *After the Ice: A Global Human History 20,000–5,000 BC*. London: Weidenfeld & Nicolson.

66. Hartog, François. 2003. *Régimes d'Historicité. Présentisme et experiences du temps*. Paris: Editions du Seuil.

67. Hartog, ibid., p. 183.

68. Evers, Adalbert and Nowotny, Helga. 1987. *Über den Umgang mit Unsicherheit. Die Entdeckung der Gestaltbarkeit von Gesellschaft*. Frankfurt am Main: Suhrkamp.

69. Nowotny, Helga. 2002. "Vergangene Zukunft: Ein Blick zurück auf die Grenzen des Wachstums." In: *Impulse geben — Wissen stiften. 40 Jahre Volkswagenstiftung*. Göttingen: Vandenhoeck & Ruprecht, pp. 655–696.

70. Ziman, John. 2000. "Selectionism and Complexity." In: Ziman, John (ed.), *Technological Innovation as an Evolutionary Process*. Cambridge: Cambridge University Press, pp. 41–51.

71. Gisler, Priska, Guggenheim, Michael, Maranta, Alessandro, Pohl, Christian, and Nowotny, Helga. 2004. *Imaginierte Laien. Die Macht der Vorstellung in wissenschaftlichen Expertisen*. Weilerswist: Velbrück Wissenschaft.

72. Giddens, Anthony. 1999. *Runaway World*. BBC Reith Lectures. Available at <http.://www.les.ac.ac.uk/giddens/lectures.htm>.

73. Pippin, Robert. 1991, 1999. *Modernism as a Philosophical Problem: On the Dissatisfactions of European High Culture*. Oxford: Blackwell.

74. Schwartz, Peter. 2003. *Inevitable Surprises: A Survival Guide for the 21st Century*. London: Free Press.

75. Beckert, Jens. 1997. *Grenzen des Marktes: die sozialen Grundlagen wirtschaftlicher Effizienz*. Frankfurt am Main: Campus.

76. Schumpeter, Joseph A. 1911, 1993. *Theorie der wirtschaftlichen Entwicklung*. Berlin: Duncker und Humblot.

77. Marx, Leo. 1997. "Technology: The Emergence of a Hazardous Concept." In: *Social Research*, vol. 64, no. 3, pp. 965–988.

78. Rosenberg, Nathan and Birdzell, L. E. 1985. *How the West Grew Rich: The Economic Transformation of the Industrial World*. New York: Basic Books, p. 265.

79. Schatz, op. cit.

80. Arthur, Brian. 1996. "Increasing Returns and the New World of Business." In: *Harvard Business Review*, July-August, pp. 1–10.

81. Perrow, Charles. 1999. *Normal Accidents: Living with High-Risk Technologies*. Princeton: Princeton University Press.

82. Bijker, Wiebe. 2005. "The Vulnerability of Technological Culture." In: Nowotny, Helga (ed.), *Cultures of Technology and the Quest for Innovation*. New York: Berghahn.

83. Helpman, Elhanan (ed.). 1998. *General Purpose Technologies*. Cambridge, Mass.: MIT Press.

84. Arthur, op. cit.

85. Hughes, P. Thomas. 1998. *Rescuing Prometheus*. New York: Pantheon Books.

86. Mokyr, Joel. 2002. *The Gifts of Athena: Historical Origins of the Knowledge Economy*. Princeton: Princeton University Press.

87. Shackle, G.L.S. 1969. *Decision, Order, and Time in Human Affairs*. Cambridge: Cambridge University Press.

88. Menger, Pierre-Michel. 2002. *Portrait de l'artiste en travailleur. Métamorphoses du capitalism*. Paris: Editions du Seuil et la République des Idées.

89. Born, Georgina. 2004. *Uncertain Vision: Birt, Dyke and the Reinvention of the BBC*. London: Secker and Warburg.

90. Green, Stephen. 2000. "Negotiation with the Future: The Culture of Modern Risk in Global Financial Markets." In: *Environment and Planning* and in: *Society & Space*, vol. 18, pp. 77–89.

Epilogue

91. MacKenzie, Donald. 2002. "The Imagined Market." In: *London Review of Books*, vol. 24, no. 21, pp. 22–24.

92. Merton, Robert C. 1992. *Continuous-Time Finance*. Cambridge, Mass.: Blackwell, p. 470.

93. Ganswindt, Thomas (ed.). 2004. *Innovation: Versprechen an die Zukunft*. Munich: Hoffmann und Campe.

94. Mirowski, Philip. 2004. *The Effortless Economy of Science?* Durham: Duke University Press; Greenberg, Daniel S. 2001. *Science, Money, and Politics: Political Triumph and Ethical Erosion*. Chicago: University of Chicago Press.

95. Erwin, Douglas H. and Krakauer, David C. 2004. "Insights into Innovation." In: *Science*, vol. 304, pp. 1117–1119.

96. Pippin, Robert. 2004. *Moral und Moderne. Die Welt von Henry James*. Munich: Wilhelm Fink Verlag. (English original: 2000. *Henry James and Modern Life*. Cambridge: Cambridge University Press.

97. Pippin, Robert. 1991, 1999. *Modernism as a Philosophical Problem: On the Dissatisfactions of European High Culture*. Oxford: Blackwell.

98. As classically narrated by Norbert Elias. 1976. *Über den Prozeß der Zivilisation. Soziogenetische und psychogenetische Untersuchungen.* Frankfurt am Main: Suhrkamp.

99. See, for example, Held, Robin. 2004. *Gene(sis): Contemporary Art Explores Human Genomics.* Available at <www.gene-sis.net>; Ackerman, Susan. Available at <www.dartmouth.edu/~religion/faculty/ackerman-bio.html>.

100. Eisenstadt, Shmuel N. (ed.). 2002. *Multiple Modernities.* New Brunswick, N.J.: Transaction.

101. Tomkins, Richard. 2004. "Men on Mars Are Just So Last Century." In: *Financial Times,* January 16, 2004, p. 10.

102. Testa, Giuseppe and Harris, John. 2004. "Ethical Aspects of ES Cell-Derived Gametes." In: *Science,* vol. 305, September 17, p. 1719.

103. McSherry, Corynne. 2001. *Who Owns Academic Work? Battling for Control of Intellectual Property.* Cambridge, Mass.: Harvard University Press.

104. Nowotny, Helga. 2005. "The Changing Nature of Public Science." In: Nowotny, Helga et al., *The Public Nature of Science under Assault: Politics, Markets, Science and the Law.* New York: Springer.

105. Shattuck, Roger. 1996. *Forbidden Knowledge: From Prometheus to Pornography.* New York: St. Martin's Press.

106. Blumenberg, Hans. 1966. *Die Legitimität der Neuzeit.* Frankfurt am Main: Suhrkamp.

107. Lefevbre, Henri. 1978. *Einführung in die Modernität. Zwölf Präludien.* Frankfurt am Main: Suhrkamp.

108. Pippin, Robert. 2004. *Moral und Moderne. Die Welt von Henry James.* Munich: Wilhelm Fink Verlag, p. 192.

109. Greenfield, Susan. 2003. *Tomorrow's People: How 21st-Century Technology Is Changing the Way We Think and Feel*. London: Allen Lane.

110. Metzinger, Thomas. 2003. "Der Begriff einer 'Bewusstseinskultur.' " In: Kaiser, G. (ed.), *Jahrbuch 2002/2003 des Wissenschaftszentrums Nordrhein-Westfalen*. Düsseldorf: Wissenschaftszentrum Nordrhein-Westfalen, pp. 150–171.

111. Kaku, Michio. 1998. *Visions: How Science Will Revolutionize the Twenty-first Century*. Oxford: Oxford University Press.

Inside Technology

edited by Wiebe E. Bijker, W. Bernard Carlson, and Trevor Pinch

Janet Abbate, *Inventing the Internet*

Atsushi Akera, *Calculating a Natural World: Scientists, Engineers and Computers during the Rise of U.S. Cold War Research*

Charles Bazerman, *The Languages of Edison's Light*

Marc Berg, *Rationalizing Medical Work: Decision-Support Techniques and Medical Practices*

Wiebe E. Bijker, *Of Bicycles, Bakelites, and Bulbs: Toward a Theory of Sociotechnical Change*

Wiebe E. Bijker and John Law, editors, *Shaping Technology/Building Society: Studies in Sociotechnical Change*

Stuart S. Blume, *Insight and Industry: On the Dynamics of Technological Change in Medicine*

Pablo J. Boczkowski, *Digitizing the News: Innovation in Online Newspapers*

Geoffrey C. Bowker, *Memory Practices in the Sciences*

Geoffrey C. Bowker, *Science on the Run: Information Management and Industrial Geophysics at Schlumberger, 1920–1940*

Geoffrey C. Bowker and Susan Leigh Star, *Sorting Things Out: Classification and Its Consequences*

Louis L. Bucciarelli, *Designing Engineers*

H. M. Collins, *Artificial Experts: Social Knowledge and Intelligent Machines*

Paul N. Edwards, *The Closed World: Computers and the Politics of Discourse in Cold War America*

Herbert Gottweis, *Governing Molecules: The Discursive Politics of Genetic Engineering in Europe and the United States*

Joshua M. Greenberg, *From Betamax to Blockbuster: Video Stores and the Invention of Movies on Video*

Kristen Haring, *Ham Radio's Technical Culture*

Gabrielle Hecht, *The Radiance of France: Nuclear Power and National Identity after World War II*

Kathryn Henderson, *On Line and On Paper: Visual Representations, Visual Culture, and Computer Graphics in Design Engineering*

Christopher R. Henke, *Cultivating Science, Harvesting Power: Science and Industrial Agriculture in California*

Christine Hine, *Systematics as Cyberscience: Computers, Change, and Continuity in Science*

Anique Hommels, *Unbuilding Cities: Obduracy in Urban Sociotechnical Change*

David Kaiser, editor, *Pedagogy and the Practice of Science: Historical and Contemporary Perspectives*

Peter Keating and Alberto Cambrosio, *Biomedical Platforms: Reproducing the Normal and the Pathological in Late-Twentieth-Century Medicine*

Eda Kranakis, *Constructing a Bridge: An Exploration of Engineering Culture, Design, and Research in Nineteenth-Century France and America*

Christophe Lécuyer, *Making Silicon Valley: Innovation and the Growth of High Tech, 1930–1970*

Pamela E. Mack, *Viewing the Earth: The Social Construction of the Landsat Satellite System*

Donald MacKenzie, *Inventing Accuracy: A Historical Sociology of Nuclear Missile Guidance*

Donald MacKenzie, *Knowing Machines: Essays on Technical Change*

Donald MacKenzie, *Mechanizing Proof: Computing, Risk, and Trust*

Donald MacKenzie, *An Engine, Not a Camera: How Financial Models Shape Markets*

Maggie Mort, *Building the Trident Network: A Study of the Enrollment of People, Knowledge, and Machines*

Peter D. Norton, *Fighting Traffic: The Dawn of the Motor Age in the American City*

Helga Nowotny, *Insatiable Curiosity: Innovation in a Fragile Future*

Nelly Oudshoorn and Trevor Pinch, editors, *How Users Matter: The Co-Construction of Users and Technology*

Shobita Parthasarathy, *Building Genetic Medicine: Breast Cancer, Technology, and the Comparative Politics of Health Care*

Paul Rosen, *Framing Production: Technology, Culture, and Change in the British Bicycle Industry*

Susanne K. Schmidt and Raymund Werle, *Coordinating Technology: Studies in the International Standardization of Telecommunications*

Wesley Shrum, Joel Genuth, and Ivan Chompalov, *Structures of Scientific Collaboration*

Charis Thompson, *Making Parents: The Ontological Choreography of Reproductive Technology*

Dominique Vinck, editor, *Everyday Engineering: An Ethnography of Design and Innovation*